Zooarchaeology of the Pleistocene/Holocene Boundary

Proceedings of a Symposium Held at the 8th
Congress of the International Council for
Archaeozoology (ICAZ)
Victoria, British Columbia, Canada
August 1998

Edited by

Jonathan C. Driver

BAR International Series 800
1999

Published in 2019 by
BAR Publishing, Oxford

BAR International Series 800

Zooarchaeology of the Pleistocene/Holocene Boundary

ISBN 9781841711096 paperback
ISBN 9781407351292 e-book

DOI https://doi.org/10.30861/9781841711096

A catalogue record for this book is available from the British Library

This book is available at www.barpublishing.com

BAR Publishing is the trading name of British Archaeological Reports (Oxford)
Ltd. British Archaeological Reports was first incorporated in 1974 to publish
the BAR Series, International and British. In 1992 Hadrian Books Ltd became
part of the BAR group. This volume was originally published by John and
Erica Hedges in conjunction with British Archaeological Reports (Oxford) Ltd /
Hadrian Books Ltd, the Series principal publisher, in 1999. This present volume
is published by BAR Publishing, 2019.

BAR
PUBLISHING

BAR titles are available from:

BAR Publishing
122 Banbury Rd, Oxford, OX2 7BP, UK
EMAIL info@barpublishing.com
PHONE +44 (0)1865 310431
FAX +44 (0)1865 316916
www.barpublishing.com

TABLE OF CONTENTS

LIST OF FIGURES

LIST OF TABLES

PREFACE: ZOOARCHAEOLOGY OF THE PLEISTOCENE/HOLOCENE BOUNDARY

Jonathan C. Driver

Department of Archaeology, Simon Fraser University, Burnaby BC V5A 1S6 Canada

The papers in this volume are the result of my participation in a symposium organized by the Working Group on the Archaeology of the Pleistocene-Holocene Transition (Chair: L.G. Straus; Secretary: B.V. Eriksen) at the XIV INQUA Congress in Berlin, 1995. At that symposium a number of detailed regional studies were presented, with a focus on human adaptation (see Eriksen and Straus 1998). A common problem for many of the regional studies was that sites of different time periods were often located in different environments or micro-environments. Furthermore, dating was often based on small numbers of radiocarbon dates or on dates inferred from artifacts. Therefore it was sometimes difficult to assess how rapid climatic change had affected human adaptation, because of problems in creating a good chronology for sites in different locations.

I proposed that another symposium be organized in which participants looked specifically at subsistence, preferably at stratified sites with a series of radiocarbon dates. I argued that by controlling site location, stratigraphy and radiocarbon age, we might get a finer level of resolution of data concerning human response to rapid environmental change. As many subsistence studies at the Pleistocene/Holocene boundary deal with animal bones, I took advantage of the 8th Congress of the International Council for Archaeozoology (ICAZ), held in Victoria, Canada, in 1998. A call for papers was sent out, asking for submissions which addressed the zooarchaeology of stratified, dated sites in the period 13,000 to 9,000 BP. Eighteen papers were scheduled for the day long symposium, of which fourteen were presented. The present volume contains one paper which was scheduled and not presented, seven papers which were presented in the symposium, one paper presented in a different ICAZ symposium, and one paper submitted after the ICAZ Congress. The papers provide a sample of zooarchaeological methods and results, but a much larger volume would be required to provide a comprehensive study of human subsistence during times of rapid environmental change.

The papers in this volume have been edited lightly to ensure similarity of format, but the content remains largely as submitted. I have cut the occasional figure or table if I felt they were redundant. I have retained regional spelling and grammar because cultural differences are important. I did not seek a grand synthesiser to make sage pronouncements. Instead, the following comments are intended to introduce some ideas which transcend the content of individual papers. The comments result from my having read these texts far too many times.

First, there is considerable variation in the way in which sites have been dated. Some sites, such as La Riera (Straus; Craighead) or Sheriden (Tankersley) are well stratified with numerous dates; others, such as Cueva del Mylodon (Borrero) have some stratigraphic problems, but radiocarbon dating can be used to define the age of fossil assemblages; in other areas dating requires a variety of methods, such as index fossils and geology (Khenzykhenova and Alexeeva); problems of contamination still plague chronologies in some areas (Driver). As demonstrated at the symposium and in the submitted papers, many archaeologists who work on the Pleistocene-Holocene transition do not have many sites in their regions which span the transition, which are finely stratified and which are dated by multiple dates on individual strata.

Second, the type and intensity of environmental change described varies from one region to another. In western Canada environments changed from uninhabitable to inhabitable. Studies of the Near East (Surovell; Munro; Dobney et al.) are less concerned with environmental change, and highlight the importance of human population increase and processes of sedentism. A future symposium might consider different types of environmental change and how human response varied.

Third, all papers stress the importance of the human/environment interaction, largely from a cultural ecological standpoint common to much environmental archaeology in general and zooarchaeology in particular. For at least the last fifteen years some archaeologists have suggested that it is possible to develop accounts and explanations of the past in non-ecological frameworks. There has been some interest in non-ecological questions by zooarchaeologists, but most zooarchaeological studies of social organization, ethnicity, gender or ideology have concerned relatively recent and relatively "complex" societies. Zooarchaeologists have been reluctant to use their data in interpretation of, or even speculation about, the way in which hunter-gatherers perceived their environment or the animals around them. An obvious future challenge for zooarchaeologists studying late Pleistocene and early Holocene faunas is to try to understand human perception of environmental change.

Acknowledgements

I am very grateful to all the authors for submitting their papers by the deadlines I set, and for trying very hard to follow the requested formats.

The original symposium was part of the 8th ICAZ Congress, organized by Becky Wigen, whom I thank for including the symposium and for adding extra presentations to it.

Producing camera-ready copy would not have been possible without the computers and printers of the Department of Archaeology, Simon Fraser University. I could not have made any of these recalcitrant machines work without the patient assistance of Shannon Wood.

LATE PLEISTOCENE AND EARLY HOLOCENE FAUNAS IN THE BAIKAL AREA

Fedora I. Khenzykhenova
Geological Institute, Ulan-Ude, Russia

Nadya V. Alexeeva
Moscow State University, Moscow

Introduction

The study of the fossil mammals in the Baikal Area is a very important key for reconstruction of human palaeoenvironment. Extensive excavations on archaeological sites have been carried out since the 1930's. N.M. Ermolova (1978) has investigated large mammal faunas from ancient settlements. During the last two decades, abundant material on small mammals has been obtained from several Late Pleistocene and Holocene archaeological sites, caves, and localities (Figure 1.1).

Late Pleistocene data on small mammals are mainly known from archaeological sites, most of them associated with river terrace complexes. Fossils have been obtained from the fluvial deposits composed of different-layered sands laden with gravel and sandy-loam. In some instances, such as the Mal'ta site, archaeological and faunal material are known from cover deposits represented by loess-like sands with gravel. Most archaeological sites are suitable for radiocarbon age-dating techniques because most contain charcoal or other datable material.

Rather significant material has been collected from caves and natural outcrops. Although their exact ages and stratigraphic occurrence are unknown, most are believed to be Holocene in age. The age of cave faunas is recognised by geological features, the degree of mineralization of bones, species composition and evolutionary level of development of voles.

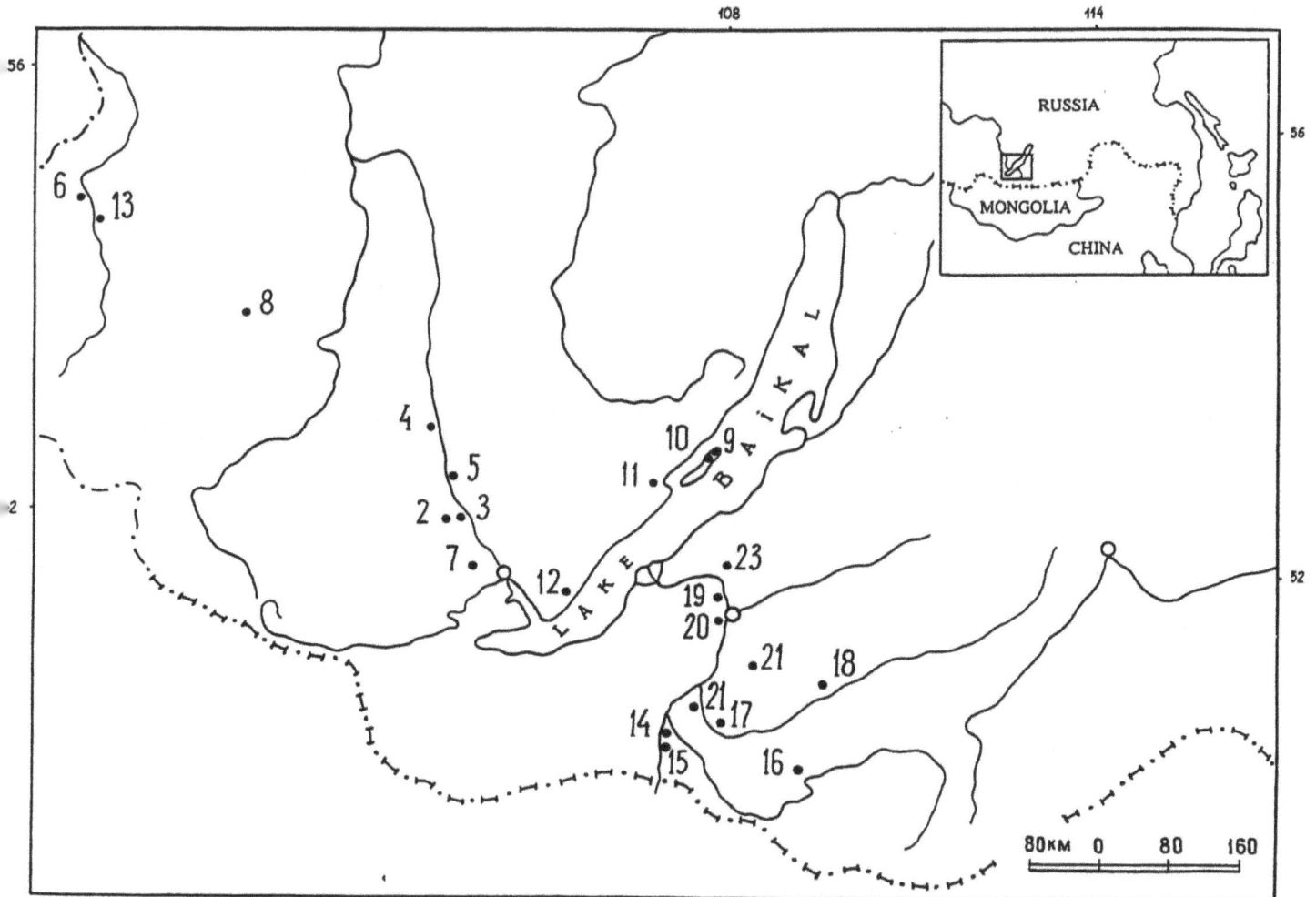

Figure 1.1. The localities of the Baikal area. 1 - Bol'shoi Jakor, 2 - Mal'ta, 3 - Ust'Belaya, 4 - Fedyaevo, 5 - Lenkovka, 6 - Ingashet, 7 - Odinskoe, 8 - Unylskaya, 9 - Usurskaya, 10 - Shamanskaya, 11 - Khurganskaya, 12 - Kadilinskaya, 13 - Koslovka, 14 - Ust'Kyakhta-17, 15 - Ust'Kyakhta-3, 16 - Studenoe-2, 17 - Kunalei, 18 - Cheremushki, 19 - Oshurkovo, 20 - Mukhino, 21 - Kharjaska-2, 22 - Chernojaravo, 23 - Cal'citovaya

Table 1.1. Age of the late Pleistocene and early Holocene faunal localities of the Baikal area

Locality	Horizon	Age	Method*	References
Fore-Baikal				
Bol'shoi Jakor	IV	11,740±140 (GIN-6461)	14C	pers. comm. E. Ineshin
	VI	12,230±250 (GIN-6466)	14C	pers. comm. E. Ineshin
	VII	12,380±250 (GIN-6467)	14C	pers. comm. E. Ineshin
	VIII	12,630±230 (GIN-6468)	14C	pers. comm. E. Ineshin
Mal'ta	V	about 12,000	geol	
Fedyaevo		about 11,000	geol	Tseytlin 1979
Lenkovka		about 8000	geol	Tseytlin 1979
Ust'Belaya		8960±60 (GIN-96)	14C	IGU 1971
Ingashet		Sr 3-4	geol	Filippov et al. 1995
Odinskoe		Sr 3-4	geol	Filippov et al. 1995
Unylskaya Cave		12,000	14C	pers. comm. L.D. Sulerzhizkyi
Khurganskaya Cave		early Holocene	paleo	Filippov et al. 1995
Shamanskaya		early Holocene	paleo	Filippov et al. 1995
Usurskaya		early Holocene	paleo	Filippov et al. 1995
Koslovka		early Holocene	paleo	Filippov et al. 1995
Kadilinskaya		early Holocene	paleo	Filippov et al. 1995
Transbaikal				
Ust'Kyakhta-17	III	11,600±155 (SO AN-3091)	14C	Tashak 1993
	IV	12,100±89 (GIN-8493a)	14C	pers. comm. L.D. Sulerzhizkyi
	V	12,230±100 (GIN-8493b)	14C	pers. comm. L.D. Sulerzhizkyi
Ust-Kyakhta-3	I	11,500	14Cgeol	pers. comm. V. Taskak
Studenoe-2	II	10,300-8000	geol	Konstantinov 1994
	III	12,700-12,000	geol	Konstantinov 1994
	IV	15,000-12,700	geol	Konstantinov 1994
Kunalei	II	10,800-10,300	geol	Bazarov et al. 1982
Cheremushki	IV	12,500-10,800	geol	Konstantinov 1994
Oshurkovo		11,230±80 (GIN-5787)	14C	Konstantinov 1994
Mukhino	VI	9500±470	14C	Yaroslavzeva 1996
Kharjaska-2	IV	Sr 3-4	paleo	Khenzykhenova 1991
	II	early Holocene	paleo	Khenzykhenova 1991
Chernojarovo		early Holocene	paleo	Khenzykhenova 1991
Cal'citovaya Cave		early Holocene	paleo	unpublished data

*geol = geological; paleo = geological age according to the evolutionary development of vole molars
** A = archaeological; C = cave; L = locality

Radiocarbon data and geological age of the faunal horizons of all localities are given in Table 1.1. The large mammal data are taken from the literature.

Late Pleistocene data on small mammals are mainly known from archaeological sites, most of them associated with river terrace complexes. Fossils have been obtained from the fluvial deposits composed of different-layered sands laden with gravel and sandy-loam. In some instances, such as the Mal'ta site, archaeological and faunal material are known from cover deposits represented by loess-like sands with gravel. Most archaeological sites are suitable for radiocarbon age-dating techniques because most contain charcoal or other datable material.

Rather significant material has been collected from caves and natural outcrops. Although their exact ages and stratigraphic occurrence are unknown, most are believed to be Holocene in age. The age of cave faunas is recognised by geological features, the degree of mineralization of bones, species composition and evolutionary level of development of voles.

Mammal faunas during the end of the Pleistocene in the Baikal area.
The Baikal area consists of two different natural zones: periglacial Siberian (fore-Baikal region) and non-glacial arid Central Asian (Transbaikalian region), and belongs to two paleozoogeographical subareas: European-Siberian and Central Asian. Different natural conditions are observed in the various species compositions of their mammal faunas during the end of Pleistocene and early Holocene.

Late Pleistocene (Sr 3-4, namely Sartan glacial period) micromammal fauna of the Fore-Baikal region was characterised by a combination of steppe, forest and tundra species and included micromammals from the B.Jakor', Mal'ta, and Fedyaevo archaeological sites; Unylskaya Cave; Ingashet and Odinskoe locations: *Lepus timidus, Ochotona*

2

hyperborea, Marmota sp., *Spermophilus* cf. *parryi, S. undulatus, Eutamias sibiricus, Cricetulus* sp., *Clethrionomys rutilus, C. rufocanus, Lagurus lagurus, Dicrostonyx* cf. *gulielmi, Lemmus sibiricus, L. amurensis, Myopus schisticolor, Alticola* sp., *Arvicola terrestris, Microtus gregalis, M. oeconomus, M. fortis, M.* ex gr. *arvalis-gregalis, M. middendorffii* (Table 1.2). The large mammal fauna of the "Upper Palaeolithic Mammoth complex" (Gromov 1948) includes *Mammuthus primigenius, Coelodonta antiquitatis, Equus caballus, Alces alces, Rangifer tarandus, Cervus elaphus, Capreolus capreolus, Bison priscus, Canis lupus, Vulpes vulpes, Alopex lagopus, Meles meles,* and *Ursus arctos* (Ermolova 1978). This fauna is ecologically mixed and has no analogues nowadays. Non-analogue micromammal faunas are known in Eurasia (Markova 1998, Smirnov 1994) and North America (Semken 1988). The mammal fauna of the Fore-Baikal Region suggests mainly a tundra-forest-steppe environment for ancient man and a temperate cold climate during the end of Upper Palaeolithic time.

A different situation is observed in the Transbaikal area, which is characterised by a more arid and continental climate.

Species composition of the Ust'-Kyakhta-17, Ust'-Kyakhta-3, Studenoe-2, Kunalei, Cheremushki, and Oshurkovo archaeological sites and Kharyaska-2 location included small mammals: *Lepus timidus, Ochotona daurica, Marmota sibirica, Spermophilus undulatus, Cricetulus barabensis, Alticola* sp., *Lagurus lagurus, Lasiopodomys brandti, Microtus gregalis, M. ungurensis,* and *Meriones* sp. (Table 1.3) and was characterised by predominance of arid steppe forms. Large mammals also belong to the "Upper Palaeolithic Mammoth complex": *Coelodonta antiquitatis, Rangifer tarandus, Cervus* cf. *elaphus, Bison* sp., *Equus caballus,* and *Spirocerus kiakhtensis* (Konstantinov 1994, Tashak 1993). Upper Palaeolithic man inhabited Transbaikalia in an environment of widespread arid steppe with small meadow and forest biotopes and an arid climate.

Mammalian faunas in the Baikal area during the Early Holocene.
The interglacial Holocene local faunas of the Baikal region contrast with those of the Sartan glacial as a result of the disappearance of previously common species. Some taxa became extinct, other were locally extirpated and live to the north or south of the Baikal area today. Thus, the

Table 1.2. List of mammals of the fore-Baikal region. See Figure 1.1 for site names

Species	Archaeological									Locality		Cave					
	1					2	4	5	3	6	7	8	11	10	9	13	12
	IV	V	VI	VII	VIII	V			III-IV								
Sorex sp.													44				+
Eptesicus nilssoni																	+
Lepus timidus		4	47	4	4	+	+										+
Lepus sp.														7	7	+	+
Ochotona hypernorea			3	2		3						+					+
Ochotona sp.													13	5	4		+
Marmota sp.			1														
Spermophilus cf. *parryi*			32		5												
Spermophilus undulatus													7	13	8		
Eutamias sibiricus												+		1			+
Cricetulus sp.														2	5		
Mus sp.																+	+
Clethrionomys rutilus			7			7					+	+	25	7		+	+
Clethrionomys rufocanus			2									+		4			+
Clethrionomys sp.											+						+
Lagurus lagurus						4						+		13	43		
Dicrostonyx cf. *guilielmi*										+	+						
Dicrostonyx sp.												+					
Lemmus sibiricus					1												
Lemmus amurensis			2	1													
Lemmus sp.												+					+
Myopus schisticolor				6		4											+
Alticola sp.												+		6	3		
Arvicola terrestris																+	
Microtus gregalis			5			36						+				+	+
Microtus oeconomus												+				+	+
Microtus fortis														3			+
M. ex gr. *arvalis-gregalis*											+						
Microtus middendorffii	11	6	11	44	1	6											
Castor fiber									+								

3

composition of the Holocene faunas is impoverished with respect to the Sartan, and is similar to the modern.

The Holocene fauna reflects more favourable environment conditions. In the fore-Baikal Region tundra species (*Dicrostonyx* cf. *gulielmi*, *Lemmus sibiricus* and *Microtus middendorffii*) were replaced by forest dwellers. The faunal composition of the Lenkovka and Ust'-Belaya archaeological sites, and that of the Khurganskaya, Shamanskaya, Usurskaya, Koslovka and Kadilinskaya Caves was: *Sorex* sp., *Eptesicus nilssoni*, *Lepus timidus*, *Ochotona hyperborea*, *Sciurus vulgaris*, *Lagurus lagurus*, *Cricetulus* sp., *Eutamias sibiricus*, *Castor fiber*, *Mus* sp., *Clethrionomys rutilus*, *C. rufocanus*, *Alticola* sp., *Arvicola terrestris*, *Lemmus* sp., *Myopus schisticolor*, *Microtus gregalis*, *M. oeconomus*, and *M. fortis* (Table 1.2). The geological age of the cave faunal horizons was judged by the degree of mineralization of bones and the level of evolutionary development of voles (Table 1.1). *Capreolus pygargus*, *Equus caballus*, *Alces alces*, *Cervus elaphus*, and *Rangifer tarandus* were widespread. This fauna indicates a forest-steppe environment and a more humid and warm climate than 12,000-10,000 years ago.

In the Transbaikalian Region the fauna was represented by steppe and forest micromammals at the Studenoe-2 and Mukhino archaeological sites, Cal'citovaya Cave and Kharyaska-2 and Chernoyarovo locations: *Sorex* sp., *Eutamias sibiricus*, *Spermophilus undulatus*, *Allactaga* sp., *Cricetulus barabensis*, *Meriones* sp., *Lagurus lagurus*, *Myopus schisticolor*, *Lasiopodomys brandti*, *Microtus*

gregalis, and *Apodemus* sp. (Table 1.3), and large mammals: *Coelodonta antiquitatis*, *Rangifer tarandus*, *Cervus elaphus*, *Moschus moschiferus*, *Capreolus pygargus*, *Equus hemionus*, *Meles meles*, *Procapra gutturosa*, *Saiga* sp., *Vulpes vulpes*. This kind of mammalian fauna was more stable than in the Fore-Baikal Region and indicates the spread of different types of steppes (arid, meadow- and forest-steppe) and a more favourable climate than during late Pleistocene.

The remains of *Mammuthus* sp. were found at the Nizhnyaya Dzhilinda (Sivakon) archaeological site on the Vitim river, and dated at 7880±80 BP (LE-1955), in the northern part of the Baikal area (Vetrov at al. 1993). Subarctic species managed to survive until the threshold of the eighth millenium, and since that time the Holocene fauna has been represented mainly by contemporary species.

Comparative analysis of the fore-Baikal and Transbaikal mammal faunas.
About 12,000-8,000 years ago the main features of micromammal faunas of the Baikal area were:

1. species diversity
2. ecological diversity
3. morphological and size diversity of species

1. Species Diversity
This period in the fore-Baikal region is characterised by the reorganisation of non-analogue micromammal associations (tundra-steppe, tundra-forest-steppe) into the recent forest-steppe associations. The peculiarity of the contemporaneous

Table 1.3. List of micromammals in Transbaikalian region. See Figure 1.1 for site names

Species	Archaeological Sites											Localities			C
	14			15	16			17	18	19	20	21		22	23
	III	IV	V		II	III	IV	II	IV		VI	IV	II		
Sorex sp.					1										12
Lepus timidus										+					
Lepus sp.												2	1		
Ochotona daurica	4	2	11	1								277	245	18	
Marmota sibirica	21		11									1			
Spermophilus undulatus	1			1			15				1	64	32	1	
Eutamias sibiricus				1											
Cricetulus barabensis												11	20	1	12
Clethrionomys rutilus															34
Clethrionomys rufocanus															13
Lagurus lagurus												1	1		
Myopus schisticolor										·			1		3
Alticola sp.	2														8
Microtus gregalis	4											8	39		
Microtus oeconomus															2
Microtus fortis					21										
Microtus ungurensis												6			3
Microtus sp.												106	34		6
Lasiopodomys brandti	4						3	71	8	5		13	14	14	
Allactaga sp.													8		
Meriones sp.												1	4		
Apodemus sp.															7

4

Transbaikalian faunas is the reduction in the distribution of arid steppe elements and the formation of forest biomes along the river valleys and north facing mountainsides. The mammalian fauna was rich in species due to different natural conditions. The same species diversity existed in the Russian Plain during glaciations and interglaciations (Markova 1998). The Late Holocene fauna is poorer than the one during the period under investigation and is very much like that of today.

The end of the Pleistocene is marked by significant extinction events that decimated large mammals of several continents, e.g. mammoth, woolly rhinoceros, buffaloes, cave lion, horse. Moreover, the recent Fore-Baikalian mammal fauna has no Arctic fox, *Spermophilus parryi*, pied lemming, brown lemming, steppe lemming, *Microtus middendorffii*. People exterminated Eurasian beaver after the 18th century (Ermolova 1978). Transbaikalian recent fauna has no *Spirocerus kiakhtensis, Equus hemionus, Saiga, Procapra gutturosa, Lagurus lagurus*. Recent natural habitats of *Ochotona daurica* and *Marmota* are smaller than during those times. *Lasiopodomys brandti* inhabits a small area in South-east Transbaikalia nowadays.

Differences.

In the Fore-Baikal Region the late Pleistocene typical micromammals were tundra- and steppe forms: pied and brown lemmings, steppe pika and steppe lemming. But during the end of Pleistocene steppe pika was replaced by northern pika, pied and brown lemmings were rare and were replaced by the Amur' and forest lemmings. Narrow-skulled vole occupied the dominant place of the steppe lemming. The populations of the Eurasian beaver, large toothed redback, northern redback and water voles have grown. As for the Transbaikalian species, the natural habitats of *Ochotona daurica*, Brandt's and narrow-skulled voles were very limited during the Holocene, but they dominated during the late Pleistocene. During the early Holocene the forest elements, *Myopus, Eutamias, Sorex* and forest voles appeared, and hydrophylic forms, *Microtus fortis, M. ungurensis* and *M. oeconomus* increased considerably.

Common features

Common species of the two natural zones of the Baikal Area are *Lepus, Marmota, Spermophilus undulatus, Cricetulus, Lagurus, Microtus gregalis, M. oeconomus* and *M. fortis*. The common features of the Fore-Baikal and Transbaikal faunas were the faunal reorganisation and transition to the forest-steppe species with hydrophylic elements.

2. Ecological Diversity

Inhabitants of different biotopes (tundra, forest, and steppe) provided the ecological diversity in the Baikal area some 12,000-8,000 years ago. Nowadays there are no analogues for them. Recent fauna of the Baikal area consists of steppe and forest forms.

Many species were forced south by the advancing glacial ice, and tundra-like environments existed at both arctic and temperate latitudes. The tundra animals were intermixed with species that today do not occur in tundra environments. Instead, many of these species today are restricted to boreal forests as well as steppe. There is no modern analogue for

these Pleistocene assemblages. Pleistocene mammal communities of the Baikal area contain modern species whose ranges do not overlap today. It is common in Late Pleistocene faunas of East Siberia to find species within contemporaneous deposits that presently inhabit tundra, steppe, and coniferous forests.

3. Morphological and Size Diversity of Species

As examples of morphological and size diversity we consider the narrow-skulled vole (a species common to the two natural zones of the Baikal Area) and Brandt's vole (a typical species of Transbaikalian Region in the period under investigation) (see below).

Systematic palaeontology.
Family: Cricetidae Fischer, 1817
Genus: Microtus Schrank, 1798
Microtus (Stenocranius) gregalis Pallas, 1779
Figure 1.2: 1, 2. Figure 1.3: 4. Figure 1.4: 4-6.

Material. The number of specimens is given in Tables 1 and 2.

Localities: the Fore-Baikal Region: Mal'ta (V), B.Jakor' (VI), Unylskaya Cave, Koslovka Cave, Kadilinskaya Cave; the Transbaikal Region: Ust'-Kyakhta-17 (III), Kharjaska-2 (II, IV).

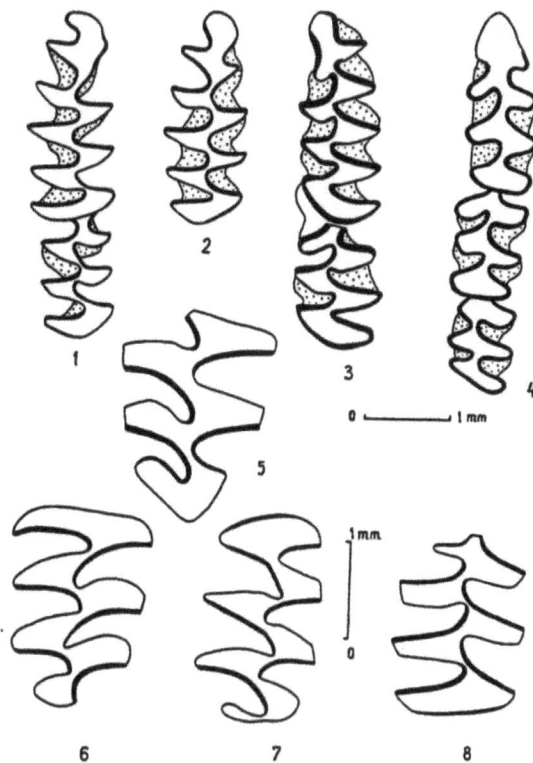

Figure 1.2. *Microtus gragalis*: 1 - M_1M_2, 2 - M_1 (Mal'ta). *Microtus middendorffiii*: : 3 - M_1M_2 (Bolshoi Jakor VI). *Clethrionomys rutilus*: 4 - M_1M_2 (Mal'ta). *Dicrostonyx cf. guilielmi*: 5 - M^3, 6 - M^1 (Ingashet), 7 - M^1, 8 - M_2 (Odinskoe)

Remarks. According to Nadachowski (1982), Markova (1986), Rekovets and Nadachowski (1995) and Dupal (1998) the evolution of M1 was directed to the complexity of the anterior part of the tooth and the increase in length from mid Pleistocene to Holocene. The shape of the anteroconid complex and development of its reentrant angles was taken as a basis for the distinction of morphotypes of M1. The molars of the narrow-skulled vole show the diversity of the morphotypes in the Baikal Area. Dimensions of M1: occlusal length 2.35-2.70-3.10 (n=23), occlusal width 0,85-1.03-1.15 mm (n=23). There is cementum in the reentrant angles. The enamel is differentiated. Predominant morphotype with some variants (Figure 1.2: 2; Figure 1.3: 4; Figure 1.4: 5-6) has the head of the anteroconid complex with two developing reentrant angles. Similar morphotype with one reentrant angle (Figure 1.4: 4) at the head of molar is very rare.

Genus: Lasiopodomys Lataste, 1887
Lasiopodomys (*Lasiopodomys*) *brandti* Radde, 1861
Figure 1.3: 1-3.

Material. The number of specimens is given in Table 1.2.

Localities: the Transbaikal Region: Ust'-Kyakhta-17 (III), Kunalei (II), Cheremushki (IV), Studenoe-2 (III, IV), Kharjaska-2 (II, IV), Chernoyarovo.

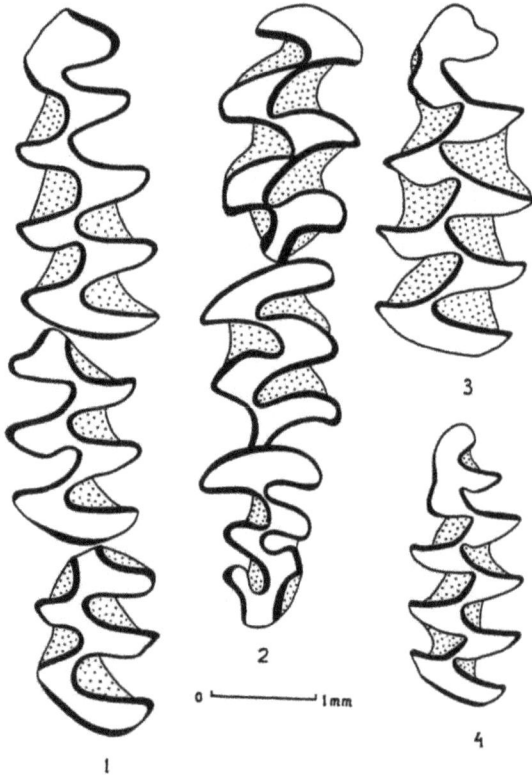

Figure 1.3. *Lasiopodomys brandti*: 1 - M_1-M_3 (Cheremushki), 2 - M^1-M^3 (Kunalei), 3 - M_1 (Chernoyarovo). *Microtus gregalis*: 4 - M_1 (Kharyaska-2 IV).

Remarks. According to M.A.Erbajeva (1976) the molars of Brandt's vole were variable and had a complex paraconid area during the Eopleistocene - recent Four morphotypes were

established. The archaic morphotype has no or only one reentrant angle at the head of the anteroconid complex in comparison with progressive ones. The material includes the archaic (Figure 1.3: 1) morphotype (5%) as well as progressive (Figure 1.3: 3) ones (95%). Dimensions of M1: occlusal length 2.65-2.95-3.30 (n=21), occlusal width 1.05-1.30-1.55 (n=21). There is cementum in the reentrant angles. The enamel is very differentiated.

Conclusions
All of these changes have culminated in the rapid emergence of new ecosystems at the end of the Pleistocene, less than 10,000 years ago. The composition, diversity, and structure of modern mammal communities are, to a large extent, the result of these environmental fluctuations.

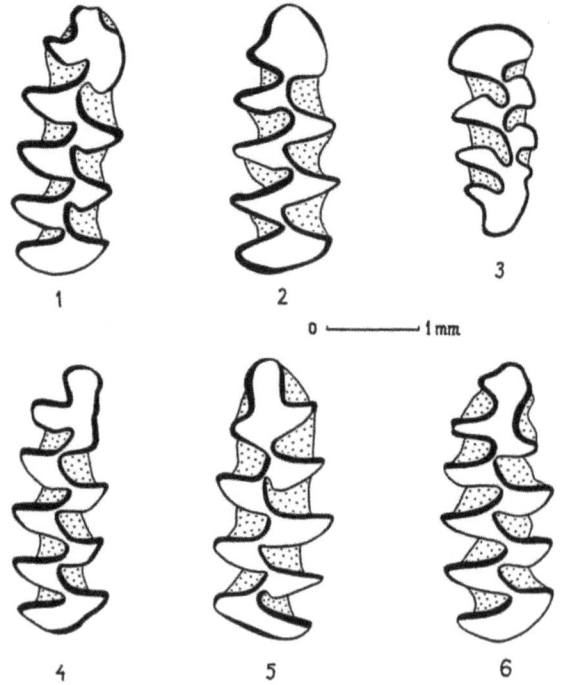

Figure 1.4. *Microtus fortis*: 1,2, - M_1, 3 - M^3 (Studenoe-2 II). *Microtus gregalis*: 4-6 - M_1 (Kharyaska-2 II).

Acknowledgements.
We would like to thank the friendly support of archaeologists: Prof. G.Medvedev, M.Konstantinov and their colleagues: E.Ineshin, E.Lipnina, L.Lbova, V.Tashak, L.Yaroslavzeva, A.Konstantinov, M.Mezherin and I.Rasgildeeva and geologists: A.Filippov and G.Shushpanova. Radiocarbon data of B.Jakor' and Ust'-Kyakhta-17 (IV-V) sites were received thanks to L.D.Sulerzhizky (Geological Institute of RAS, Moscow). Sincerely acknowledgements to them. The study of micromammalian faunas has been supported by the Russian Foundation of the Fundamental Investigations, current Grants N 97-04-49154, N 97-05-96520, N 98-04-48933, and N 98-06-80337, and Integration Program of Fundamental Research of the Siberian Branch of the Russian Academy of Sciences on Changes of the Climate and Environment in Siberia in Holocene and Pleistocene in the Context of Global Change.

References

Dupal, T.A., 1998. Evolutionary changes of size of the first lower molar in the Lineage from *Microtus* (*Terricola*) *hintoni* to the recent forms of *M.* (*Stenocranius*) *gregalis* (Rodentia, Cricetidae). *Journal of Paleontology*, 4, 87-94. (In Russian).

Erbajeva, M.A., 1976. The origin, evolution and intraspecific variability of Brandt's vole from the Anthropogene of Western Transbaikalia. *Proceedings of the Zoological Institute of Academy of Sciences of the USSR. Leningrad*, 66, 107-116. (In Russian).

Ermolova, N.M., 1978. *Theriofauna of the Valley of Angara River in Late Anthropogene*. Novosibirsk: Nauka. (In Russian).

Filippov, A.G., M.A.Erbajeva and F.I.Khenzykhenova, 1995. *Upper Cenozoic Small Mammals of South-East Siberia and its Implication for Stratigraphy*. Irkutsk: Vostsibnniiggimz Survey. (In Russian).

Gromov, V.I., 1948. Paleontologic and archaeologic foundation of the stratigraphy of Quaternary continental deposits on the SU territory. *Proceedings of Geological Institute of Academy of Sciences of USSR*, 64 (17). (In Russian).

Khenzykhenova, F.I., A.S.Endrikhinsky and M.I.Dergausova, 1991. Geology and fauna of the Kharjaska and Chernoyarovo localities. In *Questions of Cenozoic geology and fauna of Prebaikalia and Transbaikalia*. Ulan-Ude, pp.103-110. (In Russian).

Khenzykhenova, F.I., 1996. Late Pleistocene small mammals from the Baikal region (Russia). *Acta zool. cracov.*, 39 (1), 229-234.

Konstantinov, M.V., 1994. *Stone Age of the Eastern part of Baikal Asia*. Ulan-Ude-Chita. (In Russian).

Markova, A.K., 1986. Morphological peculiarities of molars in voles of the genera *Microtus, Arvicola, Lagurus*, and *Eolagurus* (Rodentia, Cricetidae) from the Mikulino-aged sites of the Russian Plain. *Proceedings of Zoological Institute of Academy of Sciences of USSR. Leningrad*, 149, 74-97. (In Russian).

Markova, A.K., 1998. Zoogeography of the Small Mammals of the Russian Plain during Recent Time. Theses of Doctor of Sciences Dissertation, Moscow: Institute of Geography of Russian Academy of Sciences, pp. 1-75.

IGU, 1971. *Mesolithic of Upper Pre-Angara region*. T.1. Irkutsk: IGU. (In Russian).

Nadachowski, A., 1982. *Late Quaternary Rodents of Poland with Special Reference to Morphotype Dentition Analysis of Voles*. Warszawa-Krakow.

Rekovets, L. and A.Nadachowski, 1995. Pleistocene voles (Arvicolidae) of the Ukraine. *Paleontologia i Evolucio*, 28-29, 145-245.

Semken, Jr., H.A., 1988. Environmental interpretations of the disharmonious Late Wisconsinian biome of South-Eastern North America. In *Late Pleistocene and Early Holocene Paleoecology and Archaeology of the Eastern Great Lakes Region*, ed. R.S.Laub, N.G.Miller and D.W.Steadman. Bulletin of Buffalo Society of Natural Sciences, 33, 185-194.

Smirnov, N.G., 1994. Rodents of Urals and Adjacent Territories in Late Pleistocene/ Holocene. Doctoral Dissertation. Ekaterinburg: Nauka. (In Russian).

Tashak, V.I., 1993. Ust'-Kyakhta-17 is the multilayered site on Selenga river region. In *Cultures and Sites of the Stones and Early Metal Epoch in Transbaikalia*. Novosibirsk: Nauka, pp.47-64. (In Russian).

Tseytlin, S.M., 1979. *Geology of Palaeolithic of North Eurasia*. Moscow: Nauka. (In Russian).

Yroslavzeva, L., 1996. Geomorphology and stratigraphy of Mukhino site. In *100th Anniversary of Hsiung-nu Archaeology. Nomadism - Past, Present in Global Context and Historical Perspective. The Phenomenon of the Hsiung-nu*. Ulan-Ude: Intern. Archaeolog. Congress Abstracts. Vol.2, pp.13-16. (In Russian).

CLIMATE CHANGE AND PATTERNS IN THE EXPLOITATION OF ECONOMIC RESOURCES (MARINE MOLLUSCA AND UNGULATE FAUNA) IN CANTABRIAN SPAIN AT THE END OF THE PLEISTOCENE, CA.21 - 6.5 KYR. BP

Alan Craighead

S40 C12 RR No.2, Lower Road, Gibsons, BC, Canada V0N 1V0

Introduction

The variation in the pattern of resource exploitation as seen in the archaeological record of Cantabrian Spain has given rise to a number of interpretations (Bailey 1978, 1983; Clark 1971, 1983; Clark and Yi 1983; Ortea 1986; Straus and Clark 1983,1986). Much of the complexity visible in the sedimentology and archaeology of cave sequences in southwestern Europe is attributable to climatic variation at the end of the last glaciation (Butzer 1981, Laville 1975, 1986). Environmental stresses in the closely coupled biogeochemical system involving the atmosphere and the earth's surface are modified and/or enhanced by changing climatic conditions (Oppenheimer 1989). The frequent, short lived, climatic oscillations during the Late Pleistocene and early Holocene had a measurable effect upon the flora and fauna available for human exploitation. This study will examine the variation in ungulate and molluscan remains in relation to climatic change during the period from about 21 to 6.5 kyr BP. The Cantabrian coast of Spain is of archaeological importance as it is one of the few regions where coastal resources may be studied over this time period which includes extreme variations in climatic conditions. The study relies on the data from the recently excavated cave site of La Riera (Straus and Clark 1986) which provides one of the most detailed and systematic analyses available in the region. La Riera has a cultural sequence extending from the early Solutrean period (ca. 20,700 BP.) through the Asturian period to about 6,500 BP., consisting of thirty stratified occupation levels, each level containing mollusc shells as well as ungulate remains. It is the only site in northern Spain which presents a detailed quantification of molluscan remains (Ortea 1986) through a sequence of occupation levels spanning the Upper Palaeolithic and the early Holocene.

Northern Spain

Climatic conditions varied extensively throughout the Late Pleistocene. Palynological analyses suggest that, during the glacial stadia, the lowlands were extensive open grasslands with a few trees in copses and small woods of pine and hardy deciduous species confined to sheltered locations (Leroi-Gourhan 1966, 1971, 1986), thus providing rich pasture for grazing animals. During interstadial conditions there was partial reforestation. A number of papers have been published which discuss the relationship between events in the North Atlantic and the vegetational, faunal and climatic history of western Europe (see for example Atkinson et al. 1987, Duplessy et al. 1981, Turner and Hannon 1988, Woillard and Mook 1982). During the last Glacial Maximum (ca.18 kyr. BP.) the polar front was pushed southward to mid- Portugal, about 40°N. Latitude (CLIMAP 1976; Ruddiman and McIntyre 1981). A steep thermal gradient immediately south of the polar front was the dominant feature, and winter temperatures north of the front were below 2°C. (McIntyre et al. 1976). This steep thermal gradient implies extremely stormy conditions along the Cantabrian coast of Spain. During the summer months the sea ice was well north of the polar front with sea surface water temperatures (SST) between 8°C. and 10°C. (CLIMAP 1976). At the time of the Glacial Maximum the mountain ranges in northern Spain were extensively glaciated (Turner and Hannon 1988). Glaciation, as indicated by terminal moraines, reached as low as 650 - 750 m above sea level (Nussbaum and Gugax 1952) and a permanent snow line possibly as low as 400 meters (Bailey 1975). Precipitation was some 33% below today's with a similar seasonal distribution (ibid.). More recent studies in the northern piedmont of the Pyrenees show that this was a period of marked aridity (Hérail et al. 1986).

It is estimated that the terrestrial ice sheets commenced melting ca. 17 kyr BP and by 16 kyr BP deglaciation was probably well advanced. With the retreat of the polar front to the north and west of Iceland the warming trend continued. ^{18}O analysis of Bay of Biscay marine cores clearly indicate the Bölling-Allerød (ca. 12.4 - 10.7 kyr BP) warming with summer SST rising some 12° - 13° C. to temperatures as warm or warmer than the present day, followed by a dramatic cooling in the Youngest Dryas (ca.10.7 - 10.1 kyr.BP.) in the order of 10° C (Duplessey et al.1981). During the Youngest Dryas the Bay of Biscay climatic conditions deteriorated rapidly to nearly full glacial conditions with summer SST falling to 11° - 12° C (Boyle 1990:48). During this phase climatic conditions were extremely variable and very stormy conditions prevailed along the Cantabrian coast with local factors magnifying or moderating events. Following the Younger Dryas, commencing about 10 kyr BP, amelioration of conditions in northern Spain is well established with substantial expansion of oak forest. Evergreen oak was present at La Moura (near Biarritz) about 9,969 BP. (Turner and Hannon 1988).

Evidence is accumulating which indicates that change in European climatic conditions was triggered by change in the mode of operation of the North Atlantic Ocean. Environmental change in western Europe is closely linked with the temperature of the northeast Atlantic (Duplessy et al. 1986). Data ($\partial^{18}O$ isotopic profiles) retrieved from the Greenland ice cores reflect and record these same events. Broecker et al. (1988:17) observe "There seems to be general agreement that the Younger Dryas impacts were synchronous in northern Europe, the northern Atlantic and Greenland". The (B-A/YD) oscillation, as recorded in the Dye 3 ice core ^{18}O profile, has been correlated with the Swiss Gerzensee Lime sediments (Dansgaard and Oeschger 1989). Duplessy et al.(1981) conclude that deglacial warming in the northeast Atlantic and the Bay of Biscay was closely correlated with that of the adjacent European mainland. Analysis of air bubbles trapped in ice from Greenland and the Antarctic indicate that the climate change associated with the Younger Dryas event was synchronous (within a few decades) over an extent of at least a hemisphere (Severinghaus et al. 1998).

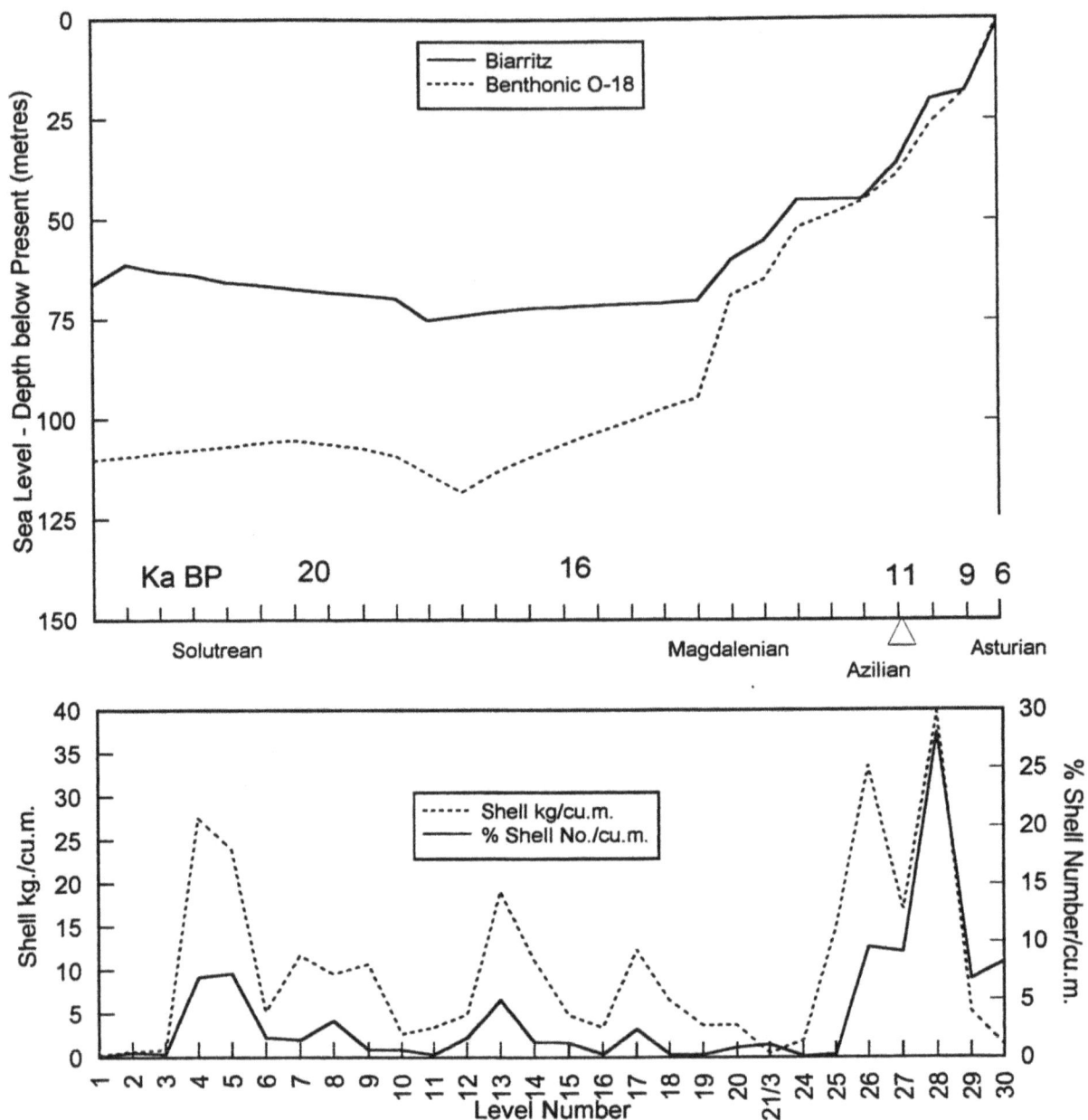

Figure 2.1. Number and weight (standardised per m³) of marine molluscan shells recovered from La Riera (bottom), plotted with sea level variation (top) for the period ca.21-6.5 kyr BP. Biarritz sea level estimates derived from Lambeck (1997); eustatic sea level estimates (benthonic ¹⁸O) from Schackleton (1987). Radiocarbon ages not to scale.

Changes in climatic conditions were extremely rapid. The Bölling - Older Dryas - Allerød - Younger Dryas (B-A/YD) oscillation was the last of a long series of similar events. Within the ice cores $\partial^{18}O$ profile values change (corresponding to 5-6° C. temperature shifts) over short time periods indicating frequent and abrupt shifts between temperate and full glacial conditions (Dansgaard and Oeschger 1989). Analysis of marine core Troll 3.1 (Lehman and Keigwin 1992) shows that throughout the B-A/YD the major SST variations occurred within periods of less than 40 years. Data from the ECM (electrical conductivity measurement) profile of ice core GISP2 (Taylor et al. 1993) indicates a climatic system that consistently and frequently oscillated between near glacial and full glacial conditions in

periods of less than 10 years, often abruptly terminating in less than 5 years.

Sea Level Change
There appears to be broad agreement that the lowering of the ocean levels along the European Atlantic at the time of the Glacial Maximum was in the range of about 100 meters with much of the subsequent rise in sea levels occurring between 15 kyr BP and 6 kyr BP (Lambeck 1997). Estimates from other locations suggest that sea levels may have been as much as 130-150 meters below modern levels (Chappell and Shackelton 1986; Chappell et al. 1994) and, after making corrections for isostatic compensation, sea shores may have been 115-125 meters below present sea levels(Fairbanks 1989; van Andel and Lianos 1984) . Recent

10

Table 2.1. La Riera *P. vulgata* radiocarbon age estimates

Level	Lab.Ref.	Age BP	$\partial^{13}C^o/_{oo}$	Adjusted age BP*	Range at 95% Probability (2r)
28	Q2933	9230±90	-0.28	8830±90	9010-8650
27 Low	Q2935	11390±70	-5.47	10990±70	11130-10850
26	Q2925	11850±85	-0.20	11450±85	11620-11280
24	Q2926	11880±75	-0.15	11480±75	11630-11330
21/3	Q2932	13200±110	-0.18	12800±110	13020-12580
18	Q2936	15990±150	-12.41	15590±150	15890-15290
17	Q2931	15760±220	-0.50	15360±220	15800-14920
14	Q2927	16810±105	-0.34	16410±105	16620-16200
7	Q2934	20400±210	-11.56	20000±210	20420-19580

estimates by Lambeck (1997), which take into account not only isostatic rebound, but also ice sheet gravitational pull, mantle viscosity and meltwater loading, suggest that in the Biarritz region the sea level was 65 to 75 meters below modern levels (Figure 2.1). This places the Cantabrian coast line at the time of the glacial maximum at a distance of 8 - 12 km. from the present day coast, correspondingly increasing the width of the coastal plain. Benthonic ^{18}O data and data from the coral reef terraces of the Huon Penninsula, Papua New Guinea (Chappell and Shackleton 1986; Chappell et al. 1994; Shackleton 1987) indicate, subsequent to 40,000 BP, that the eustatic sea level dropped to ca. -62 m below present level, rising to about -46 m about 28,000 BP, before declining to a low of -130 m during the Glacial Maximum. Based upon Lambeck's data for Biarritz, the 28,000 BP Cantabrian sea-level would be about 30 m below the present level and place the shoreline only some 3.5 km from the present coast, or approximately 5 km from La Riera cave.

Radiocarbon Dating

As previously mentioned the La Riera cave site has 30 occupation levels. The site excavators reported 28 radiocarbon age estimates, determined by five different laboratories, on charcoal and bone recovered from 15 of the stratigraphic levels (Straus 1986: Table 2.2). However, there are inversions among the determinations made on both bone and charcoal. There may have been vertical movement of the

small bone splinters and charcoal fragments used for radiocarbon assay due to human occupation or other agencies and/or the lack of sterile layers between occupation levels. In an effort to resolve some of these problems nine additional radiocarbon age estimates were obtained using *Patella vulgata* shell (Craighead 1995). These age estimates were reduced by 400 years to adjust for marine reservoir effect for apparent age (Table 2.1). It was felt that due to the large size of the limpet shells they would be less subject to migration.

A novel method of sample preparation was used to remove diagenic (recrystallized) carbonate; a grinding process was utilized rather than the conventional approach of acid etching. Several shells from each dated level were used in the preparation of the sample for radiocarbon aging as it is believed that this would provide an average age for the level. These nine estimates, all with uncertainties of less than ±250 years, were combined with ages from the literature which also had standard deviations of less than ±250 years to form a trend line.

Figure 2.2 plots the age estimates from the literature (without arrows) and the age estimates from *P. vulgata* (arrows). The solid line indicates the trend line developed from age estimates with uncertainties of 250 years or less. The less precise age determinations, with uncertainties

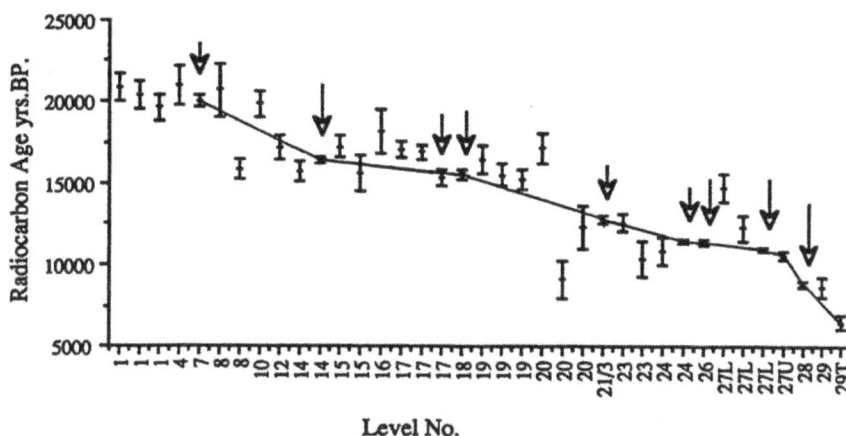

Figure 2.2. Plot of 14C age determinations from both *P. vulgata* (arrows) and the literature. Solid line: trend line developed from age estimates with uncertainties of 250 years or less. Error bars represent age ranges at the 95% probability level.

11

greater than 250 years, were examined to ascertain if they might be useful in developing a viable chronology. The radiocarbon ages used in the chronology are listed in Table 2.2. Full details of the dating methods used in the development of this chronology are presented in Craighead and Switsur (in prep.).

The age estimates in this chronology, along with sediment (Laville 1986) and pollen (Leroi-Gourhan 1986) analyses, imply that the La Riera cave site was occupied principally during the warmer climatic episodes. Late Aurignacian and earliest Solutrean occupations (levels 1 - 4) appear to have occurred towards the close of the Laugerie interstadial. The cave appears to have been most intensely occupied during the Lascaux interstadial commencing with level 12 through level 20 (ca.16.5 - 14 kyr BP). This was a period of warm climatic conditions becoming somewhat cooler and dryer in levels 19 and 20. Sediment analysis suggests that level 15 was the Lascaux optimum whereas A/C Ratio analysis (discussed below) implies the optimum occurred in level 16. This period of occupation includes the latter half of the Solutrean period, the Late Solutrean and the Early Magdalenian. The intensity of occupation during the Lascaux is indicated by the quantities of lithic tools (52% of the total from all levels) and bone/antler artifacts (50% of the total) recovered from these 9 levels. Another period of intensive occupation occurs during the Bölling-Allerød (Upper Magdalenian and early Azilian, levels 23 - 27). Asturian occupations (levels 29 - 30) did not occur until the Preboreal.

Table 2.2. Radiocarbon age estimates used for chronology building

Level No.	Lab. Ref.	14C age BP
29 Top	GaK 3046	6500±200
29	GaK 2909	8650±300
28	Q 2933	8830±90
27 Up	BM 1494	10630±120
27 Low	Q 2935	10990±70
26	Q 2925	11450±85
24	Q2926	11480±75
23	UCR 1274D	12620±300
21/3	Q 2932	12800±110
19	Q2116	15230±300
19	Q 2110	15520±350
18	Q 2936	15590±150
17	Q 2931	15360±220
15	GaK 6449	15600±570
14	UCR 1271A	15690±310
14	Q 2927	16410±105
12	GaK 6446	17210±350
10	GaK 6447	19820±390
7	Q 2934	20000±210
4	GaK 6984	20970±620
1	Ly 1783	20360±450
1	BM 1739	20860±410

Sea Surface Temperature Estimates

Vader (1975) comments that limpets appear to be unusually suitable as an indicator species for monitoring change in coastal waters. Cohen and her colleagues (Cohen and Branch 1992; Cohen et al. 1992) have shown that the environmentally controlled variation in the shell structure and mineralogy of marine mollusca can be a useful tool in making palaeoenvironmental estimates of early coastal environments. In the northern Atlantic there is a correlation between the shell structure and mineralogy of the limpet *Patella vulgata* and the SST. This relationship was tested on 224 modern *P. vulgata* shells collected from eight locations around European and United Kingdom coastlines. The relevant SST data was obtained for each site. It was found that the interior surface of the shell was composed of six separate, sometimes overlapping, bands of calcite and aragonite. These bands are described and defined by their relationship to the m band, the myostracum or muscle attachment scar (Figure 2.3). The innermost band (m-2) is composed of calcite. The m-1 band, the myostracum (m) and band m+1 are all composed of aragonite, while the two outer bands (m+2 and m+3) are calcitic. It was found that there is a relationship between the SST and the ratio of the width of the aragonite bands (m plus m+1) to the width of the outer calcite bands (m+2 plus m+3). This ratio is referred to as the aragonite/calcite ratio (A/C Ratio). For a detailed explanation of the method see Craighead (1995). The results produce a strong correlation between the A/C Ratio and the mean SST (r = 0.96).

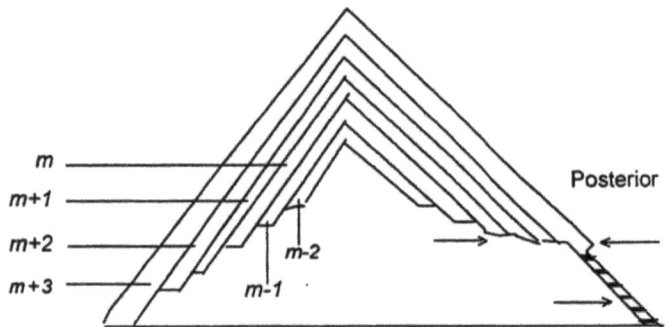

Figure 2.3. Diagrammatic sketch of a limpet shell, showing shell construction and bands (m, m+1 etc.). Arrows indicate where the shell was ground away before sampling. Sampling locations for ^{18}O analysis are indicated by the dark bands.

The relationship between SST and A/C ratio is expressed by the formula:

$$SST°C = Exponent [1.83 (Ln A/C Ratio) + 2.65]$$

This formula was then applied to 396 *P. vulgata* shells drawn from 21 of the La Riera statigraphic levels. Up to four measurements were taken on each shell and the mean A/C Ratio for each of the 21 stratigraphic levels calculated (Table 2.3). As a check on the results, 19 specimens from 5 levels (4 shells from levels 14 through 17 and 3 shells from level 18) were selected for oxygen isotope analysis. Recrystallized calcite, and all aragonite material, was removed from the edges and from about 1 cm. at the posterior end of the

Table 2.3. A/C ratio, estimated SST, and comparable geographical locations

La Riera Level	A/C Ratio	Estimated SST °C	Comparable Coastal Location	Estimated SST °C*	Comparable Coastal Location
30.1	0.72	8.8	Scotland, North	7.7	Norway, Central
29	0.96	15.8	Portugal, West	13.1	France, Brittany
28	0.69	8.6	Norway, Central	7.2	Norway, N.W.
27	0.74	9.6	Scotland, N.W.	8.2	Norway, Central
26	0.68	8.2	Norway, Central	7.0	Norway, N.W.
21/3	0.73	9.2	Scotland, North	8.0	Norway, Central
19	0.79	10.3	Isle of Man	9.2	Scotland, North
18	0.86	12.2	England, Devon	10.5	Isle of Man
17	0.89	12.9	France, Brittany	11.4	Wales, North
16	0.97	15.2	Portugal, West	13.4	France, Bordeaux
15	0.81	11.0	Wales, North/Isle of Man	9.6	Scotland, West
14	0.89	12.9	France, Brittany	11.4	Wales, North
13	0.93	13.9	France, Bordeaux	12.4	England, Devon
12	0.88	12.6	France, Brittany	11.2	Wales, North
11	0.89	12.9	France, Brittany	11.4	Wales, North
10	0.83	11.4	Wales, North	10.1	Scotland, West
9	0.84	11.7	Wales, North	10.3	Scotland, West
8	0.81	10.7	Isle of Man	9.6	Scotland, West
7	0.85	12.4	England, Devon	10.5	Isle of Man
6	0.83	11.7	Wales, North	10.1	Scotland, West
5	1.14	21.3	Morocco, West	18.0	Portugal, Southwest

*SST estimate assuming 0.5 mm loss of material from the shell edge due to taphonomic processes
Comparable location based on data from USSR Navy Charts (1979) 51-62 (Surface Air Temperature) and 128-139 (Surface Water Temperature)

bisected shells by grinding. Up to 8 samples were taken along the axis of growth at about 2.0 mm. intervals to cover at least one years growth as illustrated in Figure 2.3. The mean $\partial^{18}O$ values were compared with the A/C Ratios from the corresponding levels and show a relatively strong negative correlation (r = -0.84). It was concluded that the A/C Ratio provides a reliable guide to palaeotemperature variation. However, too much reliance should not be placed on the absolute values of the estimated SST. Many of the archaeological specimens have abraded edges and the loss of a few mm will result in an overestimate of temperature by as much as 1-2 °C. If, for example, it is assumed that the specimens used in this study had lost 0.5 mm from the shell edges through taphonomic processes, adjustment for this loss results in an average decrease in the estimated SST of 1.5°C, equivalent to a northward shift in geographical location of some 4° latitude (Table 2.3). These lower A/C Ratios and SST estimates are used in the following discussions.

Some of the absolute values appear to be questionable, notably the anomalously high value for level 5 which is outside the range of any other climatic indicators for this period. However, the presence of the limpet *Patella ruistica* and the land snails *Oestophorella bouvinieri* and *Cochlosstoma berilloni* (Seddon, personal communication) in level 4 does suggest conditions somewhat similar to the present day. Level 16 is associated with the middle of the Lascaux interstadial and a relatively high value is not unreasonable but level 29 appears to be too low in relation to level 16. Other evidence pertaining to level 29, such as

the presence of the warm water molluscs *Monodonta lineata* and Patella depressa and the almost complete absence of the cold water species *Littorina littorea*, implies temperatures as warm or warmer than the present day.

Molluscan Remains
General Trends.
Throughout the Mousterian in southwest Europe there appears to be a virtual absence of mollusc shells in the archaeological record. The lack of shells in the record prior to the Holocene does not necessarily indicate marine resources were not utilized. The archaeological invisibility of marine exploitation may well be the consequence of the melting ice caps and the glacio-eustatic sea level rise in the post Pleistocene.

Shell middens (*concheros*) have been reported on Spain's northern coast since before the turn of the century and are ascribed to the Asturian period, but many have been destroyed by the removal of shell, erosion or by excavation. One of the largest of the deposits, approximately 12 m. high by 50 m. long (Obermaier 1925:386), was estimated to contain some one million shells (Bailey 1975). None of the deposits associated with earlier cultural periods, for example the Magdalenian levels at La Riera and El Juyo (Barandiarán et al. 1985), appear to contain anything close to this quantity. The accumulation of mollusc shells in middens during the Asturian period exceeds the Magdalenian and earlier periods by several orders of magnitude. In the La Riera sequence, level 29, together with the stalagmitic crust of level 30, represents the Asturian midden with dense masses of partially cemented shells. The excavators have noted that very little of levels 29 and 30 was sampled. Figure 2.1 plots the weight and numbers of shell per cubic meter recovered from each occupation level and presents these data along with sea-level curves. Volume of excavated sediments and weight per cubic metre of shell is provided in Straus et al. (1986: Table 9.3) while standardised numbers of shells per level were obtained by combining volume data with the data provided in Ortea (1986: Table 15.2). There is a sharp increase in the quantity of shells in the Late Azilian/Asturian levels commencing about 11.5 kyr BP. Using the Biarritz data (Lambeck 1997), this would correspond to a sea level some 40 m below the present level and shoreline some 5 to 6 km distant from La Riera (Straus and Clark 1986: Figure 1.5). By the beginning of the Asturian period, ca.9 kyr BP, the sea-level would have been about 20 m below present and the distance to the shoreline only some 3.5 km. It is evident that the greatly increased quantities of shells, culminating with the Azilian/Asturian middens, are closely correlated with the early postglacial sea-level rise. In view of the nature and size of other Asturian shell deposits and the restricted nature of the sampling, it is probable that the weight and number of shells in the La Riera Asturian are considerably under-represented due to the cemented nature of the deposits and the previous removal of shell.

If the proximity of the contemporaneous shoreline is a significant factor in the development of shell deposits we should expect to find significantly increased quantities of shells at times of high sea-level stands. It is unfortunate that the La Riera sequence does not extend back to the period of moderately high sea levels of about 28 kyr BP and earlier. There is a significant increase in the quantities of shell in levels 4 and 5, dated to about 20 kyr BP, and other Solutrean and Magdalenian sites (notably El Juyo) at this time period. This accords with the implied warmer SST but occurs at a time when other evidence indicates sea levels were close to maximum regression. Reduction of the travel distance between the cave site and the shore cannot be invoked to

explain the increased representation of shells during this time period. However, it is worthy of note that the increase in the quantity of mollusca shell in La Riera levels 4 and 5, and other Cantabrian sites of this time period, coincides with a slightly higher sea-level stand estimated from benthonic ^{18}O data (Shackleton 1987). The presence in level 4 of *P. rustica* and the land snails *Oestophorella boubinieri* and *Cochlosstoma berilloni* imply warmer climatic conditions (possibly similar to the present day) and corresponds with the end of the Laugerie interstadial.

Although there is only a small decrease in the standardised numbers of shells in level 27, there is a decline in the quantity of shell per cubic meter. Although level 27 was the thickest stratum, had the largest volume of excavated sediments and the largest quantity of molluscan shell, in terms of kilograms per cubic meter, only relatively smaller quantities of shell, lithics and bone were recovered. The sparseness of archaeological material in relation to the depth and volume of the deposits suggests that some refuse may have been dumped elsewhere. Possibly, as Straus (1979) suggests for the Asturian, material was discarded on a living floor outside the entrance to the cave which, because of poor preservation, is archaeologically invisible.

Species Representation.
Ortea (1986) conducted an intensive study of the La Riera molluscan material. Four species make up the bulk of the collection; the limpets *Patella vulgata* Linnaeus,1758 and *Patella depressa* Pennant, 1777(=Patella intermedia Murray, 1758 in Ortea), the periwinkle *Littorina littorea* (Linnaeus, 1758), and the topshell *Monodonta lineata* (da Costa,1778).Through most of the sequence *P. vulgata* is dominant with small numbers of L. littorea (Figure 2.4). Commencing with level 19 there is a rise in the numbers of *P. depressa* and *M. lineata*.In the Azilian and Austurian levels (levels 27-30) the previously dominant species are almost completely replaced.The disappearance of *L. littorea*,

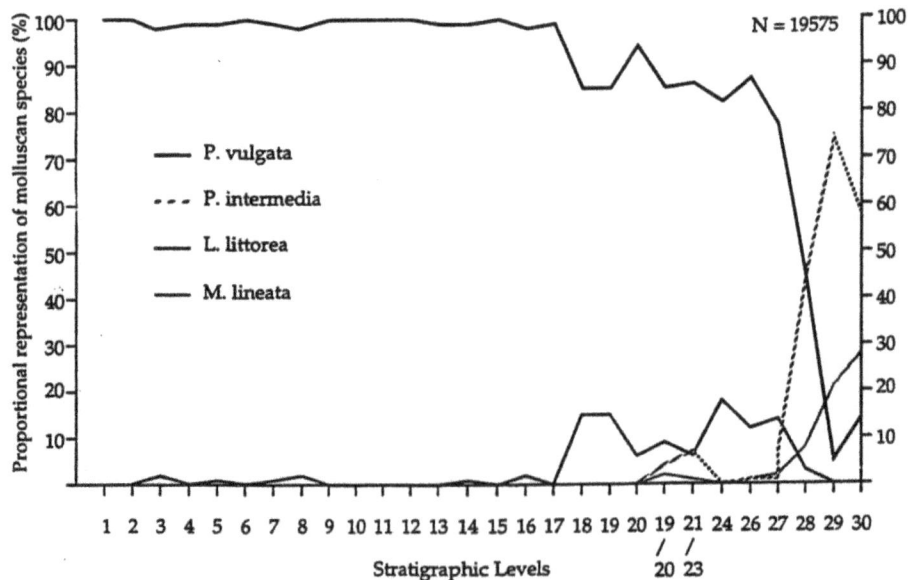

Figure 2.4. Percentage change in the proportions of edible molluscan species in the La Riera sequence, after Ortea (1986).

a northern species (Hayward and Ryland 1990), and its replacement by *M. lineata* has long been recognized as the result of the post- glacial increase in SST and the complete absence of *M. lineata* in level 30 has been cited as evidence that the SST was warmer than at present, as on the modern Cantabrian shore both species are present (Clark 1971).

Patella rustica Linnaeus, 1758, a warm water species with a range that extends from the Atlantic coast of Africa only as far north as Biarritz (Campbell 1976, Fischer-Piette and Gaillard 1959), is present only in very small numbers. Ortea (1986: table 15.3) identified 2 specimens in level 4, and an examination of 883 limpet shells, sent earlier to Cambridge for ^{18}O analysis, resulted in the identification one *P. rustica* in each of levels 21/3 and 29 (Craighead 1995). These changes in species representation confirm a climatic warming trend towards the upper levels of the La Riera sequence.

The dominance of *P. depressa* in the upper levels can similarly be explained by a warming of the SST as it has a more southerly distribution than *P. vulgata*. Its range extends along the Atlantic coast from northern Africa to the coast of north Wales, whereas *P. vulgata* ranges from northern Norway south to the Portuguese coast (Hayward and Ryland 1990). *P. depressa* prefers exposed rocky shores (Evans 1947), inhabits a restricted vertical zonation, is not found below the spring tide mean low water mark, and is most abundant in the range from mean low tide to the mean neap tide high (Southward and Orton 1954; Fischer-Piette 1948). Unlike *P. vulgata* it cannot tolerate low salinities or locations very high in the intertidal zone (Bowman and Lewis 1977; Fischer-Piette 1948; Fretter and Graham 1976). Consequently, where the two species are in competition *P. vulgata* tends to occupy the high shore and estuaries. In Portugal, close to the southern limit of *P. vulgata*, *P. depressa* replaces *P. vulgata* as the dominant species on both exposed and sheltered shores (Guerra and Guadencio 1986). It is unlikely that *P. depressa*,being the more southerly species, would have been present in any large number on the Cantabrian shores during the period of the Glacial Maximum, or during the Younger Dryas, as the water temperatures would have been too cold. By Asturian times (level 29) the SST had increased sufficiently to allow, not only *P. depressa*, but also *P. rustica* to compete with *P. vulgata*.The virtual disappearance of *P. vulgata* may also be related to glacial melting and heavy river run-off at the commencement of the Holocene when rivers deposited enormous quantities of sediments in the coastal environments. The estuarine habitats of *P. vulgata* would have been loaded with sediments and *P. vulgata* cannot feed in locations where the substrata is silted, nor survive in waters with a heavy sediment load as their primitive gill system quickly becomes clogged (Yonge and Thompson 1976:5), nor in locations where the suspended sediments restrict the penetration of light needed to promote the growth of algal vegetation (Boyle 1981:21). These factors, together with independent evidence for temperatures at least as warm as the present, provide a case for attributing the change in molluscan species to climatic warming.

An alternative explanation is the demographic hypothesis (Clark and Straus 1986; Ortea 1986). The increased quantities of shell and the almost complete replacement of *P.*

vulgata by *P. depressa* in the La Riera upper levels, and a concomitant decrease in limpet size, is seen as supporting evidence. In cost-benefit terms *P. depressa* is a less desirable resource for human exploitation than *P. vulgata* as it is generally smaller and its preferred habitat is the exposed rocky (and presumably less accessible) shores and, therefore, may be ignored until such time as the rising population places pressure on the resource base. This argument implies that *P. depressa* were available throughout the sequence but were not collected because of the lack of population pressure. The hypothesis fails if, due to environmental conditions prior to the Late Azilian/Asturian, *P. depressa* were not present on the shore and/or if the disappearance of *P. vulgata* from the record is due to climatic change. It seems unlikely that a human population, under pressure due to overpopulation to increase food supplies, would not exploit available *P. vulgata* unless it is assumed that overexploitation had almost completely wiped out the species.

Other explanations may be changes in technology or the possibility that there was a significant alteration in coastal morphology subsequent to the Younger Dryas. A change in lithic technology is visible in the Asturian. Heavy duty quartzite tools, chopping tools, choppers, bifaces and picks, are more common. The pick is considered as the index fossil for the Asturian but, to date, neither this tool nor the other heavy duty lithics have been linked with the increase in molluscan exploitation. It would appear that there has been very little change in coastal morphology as, depending upon location, both the 100 m contour line and the 100 m isobath lie approximately equidistant from the modern shoreline. Judging from the underwater contours, inlets may have been located in essentially the same locations as they are found today (Straus 1979). The present shoreline is exposed, with rocky coasts, and supports abundant populations of both principal limpet species. It is difficult to test these hypotheses by means of independent data and therefore they cannot be entirely excluded. The population hypothesis is testable, in principle, because overexploitation should result in a reduction in the size of the *P. vulgata* gathered for human consumption and this change should be visible in the archaeological record.

Size Variation.

There are at least seven principal factors which determine the growth of limpets: water temperature, exposure, zonation, salinity, food resources, inter and intra species competition, and predation, whether human or otherwise. *P. vulgata* in cold waters have a reduced growth rate or cease growing completely during winter months (Choquet 1968; Fretter and Graham 1976:27; Baxter 1982). In Cantabrian Spain where *P. vulgata* is approaching the southern limits of its range, the maximum length of specimens from similar habitats appears to be 5-10 mm. less than in Britain (Lewis 1986). *P. vulgata* on exposed shores do not grow as large as animals in sheltered locations and *P. vulgata* on the high shore or in estuaries where salinity varies also have reduced growth patterns (Ebling et. al. 1962; Fretter and Graham 1976:27; Nelson-Smith 1977). Wolcott (1973) notes that mid-shore animals have to compete with low shore species in the zone below and the high shore species in the higher zone. The impact of these well established variables complicates the interpretation of size variation in molluscan shells recovered from archaeological strata. The tendency of human predators to collect the larger specimens with a preference for limpets collected from the lower tidal zones (Mellars 1978) biases archaeological collections. Then, of course, measurement error and sampling bias are additional sources of uncertainty.

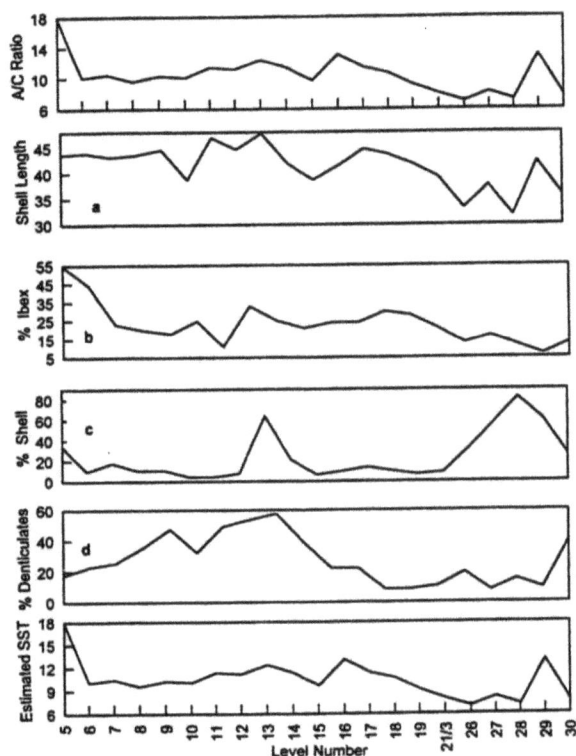

Figure 2.5. A/C Ratio and SST plotted with: (a) *Patella vulgata* shell length, (b) percentage of ibex (MNI), (c) percentage of shell weight relative to bone weight, (d) percentage of denticulated tools.

The metrical information obtained by Ortea (1986) on the La Riera limpet remains provides an extremely comprehensive data base. In all, he identified nearly 20,000 molluscan specimens and took measurements on some 18,000 limpets including 15,363 *P. vulgata*. The number of *P. vulgata* measured by Ortea includes not only complete specimens but he also estimates shell length from broken specimens based upon the thickness of the apex. Unfortunately he does not provide the formula for these estimates nor the number of specimens so estimated. In an earlier study (Craighead 1995) an attempt was made to replicate his data. New measurements were taken on 1610 *P. vulgata* shells from the original collection stored in the Oviedo Museum to which were added 518 shells stored in Cambridge. Measurements using vernier calipers, to the closest 0.5 mm, were taken only on intact shells. Comparison of the means of both samples show no significant statistical difference. Calculation of Student's t gives a value of t=0.85 which is less than the significant value of t=2.056 (using a 2 tailed test at the 0.05 significance level with 26 degrees of freedom). There are some notable differences between the two data sets, the most obvious being level 29 where the 1995 study indicates a mean length of 42.4 mm. whereas Ortea's

data give a mean length of 26.8 mm. As level 29 is critical to the interpretation of the sequence, it is most unfortunate that only 5 complete specimens were available for measurement (Ortea records data from 34 specimens). The small sample sizes compromise the evaluation of the difference in means. A smaller mean size for *P. vulgata* shell might be expected in level 30 if the SST at that time was indeed warmer than the present day (Clark 1971). As noted above, the size of *P. vulgata* tends to decrease in the warm waters near the southern edge of its range.

Palaeotemperature Variation and Changes in Shell Size
The method of estimating the SST using the A/C Ratio method was discussed above. Figure 2.5a shows the mean *P. vulgata* size compared with the A/C Ratio and indicates that the two variables track each other, the direction of change being the same for all levels with the exception of levels 5, 7 and 17. The decrease in mean length at level 10 and the variation through to level 16 and through levels 18 to 30 are matched by variations in the A/C Ratio. If the level 5 outlier is excluded, the correlation coefficient (r = 0.76) indicates that the two variables are quite strongly correlated and the mean size of *P. vulgata* was, in turn, dependent upon the SST.

Ungulate Remains
Altuna (1986) carried out an in depth analysis of the La Riera faunal remains. Data from his study indicate, based upon MNI count, that red deer (*Cervus elaphus*) was the most important ungulate resource throughout the occupation of the cave subsequent to the early levels where horse (*Equus ferus*) and bovines (*Bos primigenius, Bison priscus*) were exploited in significant numbers. After the Early Solutrean (level 6) these two species decline in importance. The percentage of red deer increases significantly through levels 5 to 11 to a peak of 72%. In the higher levels the percentage does not drop below 50% except in level 26. When the red deer MNI percentage is compared with the A/C Ratio by level there are some similarities in the curves but regression analysis does not indicate a correlation of any significance.

In terms of MNI count, ibex (*Capra pyrenaica*) was the second most important species exploited at La Riera. Like red deer, ibex was hunted in every level. Figure 2.5b plots the MNI percentage and the A/C Ratio for the levels under study. Regression analysis shows a significant, although weak, correlation (r = 0.59). However, if small sample counts (MNI <5) are omitted the correlation coefficient increases to r= 0.82 suggesting that ibex exploitation occurred during periods of warmer climatic conditions.

Shell/Bone - A/C Ratio
From level to level there is a large variation in the weight of shell relative to the weight of recovered bone. If there was a warm climate preference in the exploitation of the marine littoral resources it would be expected that the percentage of shell remains (as a % of the bone weight) would fluctuate in concert with the A/C Ratio. Regression analysis indicates a significant, although weak correlation (r= 0.57). However, the relationship between the A/C Ratio and the shell percentage may be stronger than indicated; the rise and fall of the shell percentage is almost synchronous with the variation in the A/C Ratio, only the degree of change is

16

different (Figure 2.5c). This analysis does suggest that in levels 5 through 19 the consumption of shellfish increased relative to ungulate consumption in periods of higher A/C Ratio and SST.

Lithics

The relationship between climate and the use of lithic tools was examined by comparing the percentage of each tool type in each level with the A/C Ratio. Rolland and Dibble (1990) have found that much of the variability observed in Middle Palaeolithic tool assemblages is related to environmental conditions. It was Bordes (as cited in Binford 1969) who first suggested that denticulates were used in the processing of plant material. If this hypothesis is correct then one would expect an increase in the presence of denticulates in levels with higher SST. A weak but statistically significant correlation (r = 0.56) was found to exist between denticulate tools and the A/C Ratio when the level 5 outlier is not included. Again, this relationship may be stronger than indicated by the coefficient as the rise and fall of the two variables track each other, rising through levels 8 to 13, declining in level 15, an increase in level 16, thence dropping to a minimum in level 26 and rising again in level 29 (Figure 2.5d).

The increased percentage of denticulated tools in levels with higher A./C Ratios gives support to the suggestion that they were used to process plant material. No correlation with climate change was observed in the variation of other tool categories.

Conclusions

The results demonstrate that the analysis and interpretation of long-term palaeoclimate and palaeo-economic trends in a period of rapidly oscillating climatic conditions, with extremes fluctuating between fully glacial and interglacial conditions, is a complex procedure. The principal outcome of the above analyses has been to show that the variation in the patterning of molluscan and ungulate remains in the La Riera sequence is most likely the result of climate and habitat changes. Change in the numbers and species of marine mollusca appears to be closely related to environmental change. In levels where the distance to the coast line is reduced due to sea-level rise, large quantities of limpets were transported to La Riera cave. Also, the limpet gatherers were less selective with regard to size with smaller specimens being transported to the cave.

Obvious associations with environmental change are the presence or absence of horse, bovids and reindeer. Change in the numbers of ibex present throughout the sequence is less obviously associated with environmental change although the data imply that the exploitation of ibex increased with warmer climatic conditions. During stadial conditions climatic conditions were not only cold but, also, arid (Hérail et al. 1986, van Campo 1984). Increased humidity and snowfall during warmer conditions may have driven the ibex to lower elevations. While ibex represents a high percentage of the MNI counts at La Riera, the species is not abundant at other Cantabrian sites. Only a total of 15 is reported for the Cueva Morín sequence (Altuna 1973), present in only two levels at Castillo (Cabrera-Valdes 1984), under 5% of the ungulate species at El Juyo (Freeman 1973) and no ibex

reported at El Pendo (Fuentes-Vedante 1980). It would appear that the large numbers of ibex exploited at La Riera is a factor of the site's location close to the foot of the coastal mountains, the Picos de Europa. Ibex would have been less accessible and/or less predictable during periods of cold climate (low SST).

The data indicate that the consumption of shellfish relative to ungulates also increased in periods of higher SST as does the percentage of denticulated tools. These data suggest that the cave occupants exploited a wider range of resources and may have enjoyed a more varied diet during the warmer climatic episodes.

The analysis of the molluscan data show that the changes in shellfish species representation are most likely the result of environmental variation. Analytical techniques based on the mineralogy and morphology (and the derived metrical data) of the exploited mollusc shells, supported by studies of modern populations, support this conclusion. Using such methods, the relative influences of palaeoenvironment, palaeoecology and palaeoeconomics can be disentangled, at least to some degree. However, it is impossible to completely eliminate the possibility that increased exploitation by human populations may have had some influence. The number of sites along the Cantabrian coast certainly increases, rising from 36 in the Upper Magdalenian to 110 in the Mesolithic (includes Azilian, Asturian and Tardenosian). The Mesolithic sites cover a period of approximately 5,000 years, or about 22 sites established per millennium (Clark and Yi 1983). This number does not necessarily imply population growth along the Cantabrian coast; the increase in sites could be the result of a more mobile life style. However, other factors cannot be excluded, particularly if Ortea's smaller mean shell size for *P. vulgata* in level 29 is preferred.

There are ambiguities in the La Riera data principally due to the inadequacy of some of the samples available for study. It is certainly desirable to examine larger samples of *Patella vulgata* shells from Asturian deposits. Nevertheless, it can be stated that the results of this study indicate that any population growth in the Cantabrian region does not appear to have had a significant impact on the availability or the growth pattern of edible limpet species.

A source of uncertainty is the nature of the Cantabrian coast at the various sea-level stands during the late Pleistocene. The morphological and ecological character of the coastline would have changed with regard to such variables as bays and river estuaries, width and nature of the substrata and the degree of exposure. This study has tended to discount these variables in the discussion of the La Riera data on the basis that a rocky shoreline with a similar underwater gradient would be less liable to major change with rising sea-levels than shallower coastal areas.

The quantities of mollusc shells discarded in the Asturian concheros far exceeds the numbers of shells found in any of the known deposits from earlier periods. The presence of mollusc shells in the cave site at La Riera and other locations confirms the exploitation of the marine littoral during the Upper Palaeolithic. It seems unlikely that all the shellfish collected were brought to the cave for consumption.

In limpet species the average wet meat weight is about 48% of the total weight (Buchanan 1988: 122). As over 50% of the limpet weight represents the weight of the shell, transportation costs suggest that the large majority of shells would have been discarded close to where they were gathered and only the larger specimens carried to the cave. In levels where the distance to the coastline was reduced due to the rise in sea-level, large quantities were transported to the La Riera site and gatherers were less selective in the size of the limpets. It is possible that just as many shellfish were being collected in the Upper Palaeolithic as in the Mesolithic, the majority discarded closer to the contemporaneous shoreline and now submerged as the result of eustatic sea-level rise; the increased visibility of shell in the Azilian/Asturian resulting from the proximity of the coast line. Whether this is strictly true has yet to be confirmed.

Acknowledgements
I am most grateful to the British Academy for funding the radiocarbon dating and the isotope analyses. For permission to study the La Riera collection in the Museum of Archaeology, Oviedo, I wish to express my thanks to the Consejera de Cultura y Educación, Principado de Asturias, Dña. Matilde Escortell Ponsoda, Director of the Museum, and the excavators Geof Clark and Lawrence Straus and also Manuel González Morales for his assistance and advice in Spain. For access to the facilities of The Godwin Laboratory, Cambridge, and for assistance with the analytical work, thanks are due to Nick Shackleton, Roy Switsur, Mike Hall and Andrew Gerrard. I thank Roy Switsur for processing and interpreting the radiocarbon assays. Thanks are also due to Steve Hawkins, Rosemary Bowman, Mike Kendall, J.D. Fish, Mick Wood, Louis Cabioch, Jean-Louis Birrien, Ole J. Lönne, Jon-Arne Sneli, Mary Seddon, Glen Jamieson, and Junkal Peña Othaitz for advice and assistance in the collection of modern *Patella vulgata* shells, relevant water temperature information and also advice on molluscan ecology. Thanks to Nick Winder and Jack Nance for advice on statistical procedures.

References

Altuna, J., 1973. Fauna de mamíferos de la Cueva de Morín. In *Cueva Morín: Excavaciones 1969*, ed. J. González Echergary and L. Freeman, eds. Santander: Publicaciones del Patronato de las Cuevas Prehistóricas de la Provincia de Santander, pp. 281-290.

Altuna, J., 1986. The mammalian faunas from the prehistoric site of La Riera. In *La Riera Cave: Stone Age Hunter-Gatherer Adaptations in Northern Spain*, ed. L.G. Straus and G.A. Clark. Arizona State University Anthropological Research Papers No. 36, pp. 237-274, 421-480.

Atkinson, T.C., K.R. Briffa and G.R. Coope, 1987. Seasonal temperatures in Britain during the last 22,000 years, reconstructed using beetle remains. *Nature* 325, 587-592.

Bailey, G.N., 1975. The Role of Shell Middens in Prehistoric Economies. Unpublished Ph.D. dissertation. University of Cambridge.

Bailey, G.N., 1978. Shell middens as indicators of postglacial economies: a territorial perspective. In *The Early Postglacial Settlement of Northern Europe*, ed. P.A. Mellars. London: Duckworth, pp. 37-63.

Bailey, G.N., 1983. Problems in site formation and the interpretation of spatial and temporal discontinuities in the distribution of coastal middens. In *Quaternary Coastlines and Marine Archaeology*, ed. P.M. Masters and N.C. Flemming. London: Academic Press, pp. 559-582.

Barandiarán, I., L.G. Freeman, J. González Eschegaray and R.G. Klein, 1985. *Excavaciones en la Cueva del Juyo*. Madrid: Centro de Investigacíon y Museo de Altamira, Monografías 14.

Bard, E., M. Arnold, P. Maurice, J. Duprat, J. Moyes J. and J.C. Duplessy, 1987. Retreat velocity of the North Atlantic polar front during the last deglaciation determined by [14]C Accelerator Mass Spectrometry. *Nature* 328, 791-794.

Baxter, J.M., 1982. Population dynamics of *Patella vulgata* in Orkney. *Netherlands Journal of Sea Research* 16, 96-104.

Binford, S.R., and L.R. Binford, 1969. Stone tools and human behavior. *Scientific American* 220, 70-84.

Bowman, R.S., and J.R. Lewis, 1977. Annual fluctuations in the recruitment of *Patella vulgata*. *Journal of the Marine Biological Association of the United Kingdom* 57, 793-815.

Boyle, K.V., 1990. *Upper Palaeolithic Faunas from South-West France*. BAR International Series 557. Oxford: British Archaeological Reports.

Boyle, P.R., 1981. *Molluscs and Man*. London: Edward Arnold.

Broecker, W.S., M. Andree, G. Bonani, W. Wolfli, H. Oeschger, G. Bonani, J. Kennett and D. Peteet, 1988. The chronology of the last deglaciation: implications to the cause of the Younger Dryas event. *Paleoceanography* 3, 1-19.

Buchanan, W.F., 1988. *Shellfish in Prehistoric Diet*. BAR International Series 455. Oxford: British Archaeological Reports.

Butzer, K.W., 1981. Cave sediments, Upper Pleistocene stratigraphy and Mousterian facies in Cantabrian Spain. *Journal of Archaeological Science* 8, 133-183.

Cabrera, V., 1984. *El Yacimiento de la Cueva de "El Castillo"*. Madrid: Consejo Superior de Investigacíones Cientificas Instituto Espanol de Prehistorica. Bibliotheca Praehistorica Hispana

Campbell, A.C., 1976. *The Hamlyn Guide to the Seashore and the Shallow Seas of Britain and Europe*. London: Hamlyn Publishing.

Chappell, J., A. Omura, M. McCulloch, T. Esat, Y. Ota and J. Pandolfi, 1994. Revised late Quaternary sea levels between 70 and 30 ka from coral terraces at Huon Peninsula. In *Study on Coral Reefs of the Huon Peninsula, Papua New Guinea*, ed. Y. Ota. Yokohama: Yokohama National University, pp. 155-171.

Chappell, J., and N.J. Shackleton, 1986. Oxygen isotopes and sea level. *Nature* 324, 137-140.

Choquet, M., 1968. Croissance et longevité de *Patella vulgata* L. (gastropode prosobranche) dans le Boulonnais. *Cahiers du Biologie Marine* 9, 449-468.

Clark, G.A., 1971. The Asturian of Cantabria: subsistence base and evidence for Pleistocene climatic shifts. *American Anthropologist* 73, 1244-1257.

Clark, G.A., 1983. Boreal phase settlement/subsistence models for Cantabrian Spain. In *Hunter-Gatherer Economy in Prehistory*, ed. G.N. Bailey. Cambridge: Cambridge University Press, pp. 96-110.

Clark, G.A. and L.G. Straus, 1986. Synthesis and conclusions - Part I. *La Riera Cave: Stone Age Hunter-Gatherer Adaptations in Northern Spain*, ed. L.G. Straus and G.A. Clark. Arizona State University Anthropological Research Papers No. 36, pp. 351-365.

Clark, G.A. and S. Yi, 1983. Niche-width variation in Cantabrian archeofaunas: a diachronic study. In *Animals and Archaeology: 1. Hunters and Their Prey*, ed. J. Clutton-Brock and C. Grigson. BAR International Series 163. Oxford: British Archaeological Reports, pp. 183- 208.

CLIMAP Project Members, 1976. The surface of the ice-age earth. *Science* 191, 1131-1137.

Cohen, A.L., and G.M. Branch, 1992. Environmentally controlled variation in the structure and mineralogy of *Patella granularis* shells from the coast of southern Africa: implications for palaeotemperature assessments. *Palaeogeography, Palaeoclimatology, Palaeoecology* 91, 49-57.

Cohen, A.L., J.E. Parkington, G.B. Brundrit and N.J. van der Merwe, 1992. A Holocene marine climate record in mollusc shells from the southwest African coast. *Quaternary Research* 38, 379-385.

Craighead, A.S., 1995. Marine Mollusca as Palaeoenvironmental and Palaeoeconomic Indicators in Cantabrian Spain. Cambridge: Unpublished Ph.D. Dissertation.

Dansgaard, W. and H. Oeschger, 1989. Past environmental long-term records from the Arctic. In *The Environmental Record in Glaciers and Ice Sheets*, ed. H. Oeschger and C.C. Langway, Jr. London: John Wiley & Sons, pp. 287-317.

Duplessy, J-C., M. Arnold, P. Maurice, E. Bard, J. Duprat and J. Moyes, 1986. Direct dating of the oxygen-isotope record of the last deglaciation by ^{14}C accelerator mass spectrometry. *Nature* 320, 350-352.

Duplessy, J-C., G. Delibrias, J.L. Turon, C. Pujol and J. Duprat, 1981. Deglacial warming of the northeastern Atlantic Ocean: correlation with the palaeoclimatic evolution of the European continent. *Palaeogeography, Palaeoclimatology, Palaeoecology* 35, 124-144.

Ebling, F.J., J.F. Sloane, J.A. Kitching and H.M. Davies, 1962. The ecology of Lough Ine XII. The distribution and characteristics of *Patella* species. *The Journal of Animal Ecology* 31, 457-470.

Evans, R.G., 1947. Studies in the biology of British limpets. Part I. The genus *Patella* in Cardigan Bay. *Proceedings of the Zoological Society of London* 117, 411-423.

Fairbanks, R.G., 1989. A 17,000-year glacio-eustatic sea level record: influence of glacial melting rates on the Younger Dryas event and deep ocean circulation. *Nature* 342, 637-642.

Fischer-Piette, E., 1948. Sur les elements de prospérité des patelles et sur leur specificité. *Journal de Conchyliologie* LXXXVIII (2), 45-96.

Fischer-Piette, E. and J.M. Gaillard, 1959. Les patelles au long des côtes Atlantiques Iberiques et Nord-Morocaines. *Journal de Conchyliologie* XCIX. 135-200.

Freeman, L.G., 1973. The significance of mammalian faunas from Paleolithic occupations in Cantabrian Spain. *American Antiquity* 38, 3-44.

Fretter, V. and A. Graham, 1976. The Prosobranch molluscs of Britain and Denmark. *Journal of Molluscan Studies*: Supplement 1.

Fuentes-Vedante, C., 1980. Estudio de la fauna de El Pendo. In *El Yacimiento de Cueva de "El Pendo"* (Excavaciones 1953-57), ed. J. González Echergaray. Madrid: Consejo Superior de Investigaciones Cientificas Instituto Espanol de Prehistorica.

Guerra, M.T., and M.J. Guadencio, 1986. Aspects of the ecology of *Patella* spp. on the Portuguese coast. *Hydrobiologia* 142, 57-69.

Hayward, P.J., and J.S. Ryland, ed.,1990. *The Marine Fauna of the British Isles and North-West Europe*. Oxford: Clarendon Press.

Hérail, G., J.Hubschman and G. Jalut, 1986. Quaternary glaciation in the French Pyrenees. In *Quaternary Glaciations in the Northern Hemisphere*, ed. V. Sibrava, D.Q. Bowen and G.M. Richmond, eds. Oxford: Pergamon Press, pp. 397-402.

Lambeck, K., 1997. Sea-level change along the French Atlantic and Channel coasts since the time of the last Glacial Maximum. *Palaeogeography, Palaeoclimatology, Palaeoecology* 129, 1-22.

Laville, H., 1975. *Climatologie et chronologie du Paléolithique en Perigord*. Etude Quaternaires, 4. Marseille: Université de Provence.

Laville, H., 1986. Stratigraphy, sedimentology and chronology of the La Riera Cave deposits. In *La Riera Cave: Stone Age Hunter-Gatherer Adaptations in Northern Spain*, ed. L.G. Straus and G.A. Clark. Arizona State University Anthropological Research Papers No. 36, pp. 25-55.

Lehman, S.J., and L.D. Keigwin, 1992. Sudden changes in North Atlantic circulation during the last deglaciation. *Nature* 356, 757-72.

Leroi-Gourhan, Arlette, 1966. Análisis polínico de la Cueva del Otero. In *Cuevo del Otero*, ed. J. González Echegaray, M. García and A. Begines. Madrid: Excavaciones Arqueológicas en España, pp. 83-85.

Leroi-Gourhan, Arlette, 1971. Análisis polínico de Cuva Morín. In *Cueva Morin Excavaciones 1966-1968*, ed. J. González Echegaray and L.G. Freeman. Santander: Publicaciones de Patronato de las Cuevas

Prehistóricas de la Provincia de Santander VI, pp. 359-365.

Leroi-Gourhan, Arlette, 1986. The palynology of La Riera Cave. In *La Riera Cave: Stone Age Hunter-Gatherer Adaptations in Northern Spain*, ed. L.G. Straus and G.A. Clark. Arizona State University Anthropological Research Papers No. 36, pp. 59-64.

Lewis, J.R., 1986. Latitudinal trends in reproduction, recruitment and population characteristics of some littoral molluscs and cirripedes. *Hydrobiologia* 142, 1-13.

McIntyre, A., N.G. Kipp, A.W.H. Be, T. Crowley, T. Kellog, J.V. Gardner, W. Prell and W.F. Ruddiman, 1976. Glacial North Atlantic 18,000 years ago: a CLIMAP reconstruction. In *Investigation of Late Quaternary Palaeoceanography*, ed. R.M. Cline and J.D. Hays. Geological Society of America Memoir 145. Boulder: Geological Society of America, pp. 43-76.

Mellars, P., 1978. Excavation and economic analysis of Mesolithic shell middens on the island of Oronsay (Inner Hebrides). In *The Early Postglacial of Northern Europe*, ed. P. Mellars. London: Gerald Duckworth.

Nelson-Smith, A., 1977. Estuaries. In *The Coastlines*, ed. R.S.K. Barnes. London: John Wiley & Sons, pp. 123-146.

Nussbaum, F. and F. Gygax, 1952. Glazialmorphologische untersuchungen im Kantabrischen Gebirge. *Jahresbericht Geographische Gesellschaft* 1951/52, 54-79.

Obermaier, H., 1925. *El Hombre Fosíl*. (2nd. edition) Comisión de Investigaciones Paleontológicas y Prehistóricas, Memoria 9.

Oppenheimer, M., 1989. Climate change and environmental pollution: physical and biological interactions. *Climatic Change* 15, 255-270.

Ortea, J., 1986. The malacology of La Riera Cave. In La Riera Cave: Stone Age Hunter-Gatherer Adaptations in Northern Spain, ed. L.G. Straus and G.A. Clark. Arizona State University Anthropological Research Papers No. 36, pp. 289-298.

Rolland, N. and H.L. Dibble, 1990. A new synthesis of Middle Paleolithic variability. *American Antiquity* 55(3), 480-499.

Ruddiman, W.F., and A. McIntyre, 1981. The North Atlantic during the last deglaciation. *Palaeogeography, Palaeoclimatology, Palaeoecology* 35, 145-214.

Severinghaus, J.P., T. Showers, E.J. Brook, R.B. Alley, and M.L. Bender, 1998. Nature 391: 141-146.

Shackleton, N.J., 1987. Oxygen isotopes, ice volume and sea level. *Quaternary Science Reviews* 6, 183-190.

Southward, A.J., and J.H. Orton, 1954. The effects of wave-action on the distribution and numbers of the commoner plants and animals living on the Plymouth breakwater. *Journal of the Marine Biological Association of the United Kingdom* 33, 1-19.

Straus, L.G., 1979. Mesolithic adaptations along the coast of northern Spain. *Quaternaria* 21, 305-327.

Straus, L.G., 1986. An overview of La Riera chronology. In *La Riera Cave: Stone Age Hunter-Gatherer Adaptations in Northern Spain*, ed. L.G. Straus and G.A. Clark. Arizona State University Anthropological Research Papers No. 36, pp. 19-21.

Straus, L.G. and G.A. Clark, 1983. Further reflections on adaptive change in Cantabrian prehistory. In *Hunter-Gatherer Economy in Prehistory*, ed. G.N. Bailey. Cambridge: Cambridge University Press, pp. 166-167.

Straus, L.G. and G.A. Clark, ed., 1986. *La Riera Cave: Stone Age Hunter-Gatherer Adaptations in Northern Spain*. Arizona State University Anthropological Research Papers No. 36.

Straus, L.G., G.A. Clark, J. Ordaz, L. Suárez and R. Esbert, 1986. Patterns of lithic raw material variation at La Riera. In *La Riera Cave: Stone Age Hunter-Gatherer Adaptations in Northern Spain*, ed. L.G. Straus and G.A. Clark. Arizona State University Anthropological Research Papers No. 36, pp. 189-208.

Taylor, K.C., G.W. Lamorey, G.A. Doyle, R.B. Alley, P.M. Grootes, P.A. Mayewski, J.W.C. White and L.K. Barlow, 1993. The 'flickering switch' of late Pleistocene climate change. *Nature* 361, 432-436.

Turner, C. and G.E. Hannon, 1988. Vegetational evidence for late Quaternary climatic changes in southwest Europe in relation to the influence of the North Atlantic Ocean. *Philosophical Transactions of the Royal Society, London B* 318, 451-485.

U.S.S.R. Ministry of Defense, Navy, 1979. *Atlas of the Oceans: Atlantic and Indian Oceans*. Oxford: Pergamon Press.

Vader, W., 1975. Range extension of *Patella vulgata* (mollusca, prosobranchia) on the island of Skjevoy, northern Norway, between 1933 and 1973. *Astarte* 8, 49-51.

van Andel, T.H. and N. Lianos, 1984. High-resolution seismic reflection profiles for the reconstruction of Postglacial transgressive shorelines: an example from Greece. *Quaternary Research* 22, 21-45.

Van Campo, M., 1984. Relations entre la végétation de l'Europe et les témperatures de surface océaniques aprés le dernier maximum glaciaire. *Pollen et Spores* 26, 453-456.

Woillard, G.M., and W.G. Mook, 1982. Carbon-14 dates at Grande Pile: correlation of land and sea chronologies. *Science* 215, 159-161.

Wolcott, T.G., 1973. Physiological ecology and intertidal zonation in limpets (Acmaea): a critical look at limiting factors. *The Biological Bulletin, The Marine Biological Laboratory, Woods Hole* 145, 389-422.

Yonge, C.M., and T.E. Thompson, 1976. *Living Marine Molluscs*. London: Collins.

HIGH RESOLUTION ARCHEOFAUNAL RECORDS ACROSS THE PLEISTOCENE HOLOCENE TRANSITION ON A TRANSECT BETWEEN 43 AND 51 DEGREES NORTH LATITUDE IN WESTERN EUROPE

Lawrence Guy Straus
Department of Anthropology University of New Mexico

Introduction

It is a truism to say that the period of the Pleistocene Holocene transition (from the beginning of Bölling, c. 13 kya uncal., until the end of Boreal, c. 8 kya uncal.) was a time of tremendous change, not only in environments, but also in resources and human adaptations worldwide. However, the purpose of the INQUA Working Group on the Archeology of the Pleistocene-Holocene Transition (P-HT) is to document and understand *interregional* variability in conditions and responses across this 5000 year span of time. In particular, the role of the Sub Working Group on High Resolution Archeofaunal Records is to survey the world for detailed evidence of changes in the exploitation of animals by humans across the transition, ideally at archeological sites for which there is fairly precise chronometric control.

The upshot of such an exercise should be the region by region demonstration of:
1.) Changes in the natural faunas (due to extirpation, extinction, biogeographic factors including the effects of inter species competition, etc.) and
2.) Changes in human subsistence practices.

Only through comparison with non-anthropogenic paleontological assemblages of similar age can one unpack these two phenomena (i.e., to separate natural presence and relative abundance of animal species from the human selection effect). Because of the abundance of relatively high-precision radiocarbon dates available for many sites in this period, the project should be able to show the rates of change both in faunal turnover and in human reactions. And it should also lead to the realistic assessment of differences among latitude belts, biozones and regions in terms of these changes. In some areas of the world these changes were abrupt and marked; in others they were gradual and more subtle. At a broader level, these inter-regional differences colored the nature of the transitions from Paleolithic (Paleoindian) to Mesolithic (Archaic) lifeways. The interactions of humans with various prey animals are at the heart of the transition in forager adaptions at the onset of the Holocene. For the first time in our evolutionary history, anatomically modern humans, with sophisticated technologies and social organizations, living at sometimes relatively high regional population densities and on all the continents (except Antarctica), faced the environmental and resource changes of a glacial-interglacial transition. In some cases, the world that their ancestors had known for millennia was brusquely terminated, but in others the changes were at most slight and slow to come. The ramifications of such differences can be significant; hence the importance of this project (see papers in Straus et al. 1996 and in Eriksen and Straus 1998).

In earlier papers (e.g., Straus 1991,1992, 1995a,b) I have proposed general comparisons among various regions of southwestern Europe (Aquitaine, Pyrenees, Vasco Cantabria and, sometimes, Portugal) across the P-HT, based on my syntheses of data of varying quality and dates of highly diverse precision. In short, these articles have maximized sample size by including both high and low resolution data to produce generalized regional overviews and contrasts. In this paper I both extend the comparison northward to include Belgium and restrict the database to a few sites which I have personally excavated and which have good radiocarbon control. On the other hand, although I have conducted research in Portugal, none of the sites I tested there had levels dating between 13 and 8 kya) (and indeed such sites with preserved fauna have yet to be excavated, studied and published in Portugal). Hence, the objective of this brief contribution is to summarize archeofaunal results from La Riera Cave (Asturias, Spain) (Straus and Clark 1986), Dufaure Rockshelter (Les Landes, France) (Straus 1995c) and Bois Laiterie Cave combined with Pape Rockshelter (Namur, Belgium) (Otte and Straus 1997; Straus et al. 1999). The data presented from La Riera are restricted to the uppermost part of the stratigraphic sequence (i.e., the Upper Magdalenian, Azilian and Asturian levels). The archeozoological research summarized here is the work of Jesus Altuna (with Koro Mariezkurrena) for La Riera and Dufaure, and of Achilles Gautier (for Bois Laiterie and Pape). Specific references are: Altuna (1986), Altuna and Mariezkurrena (1995), Gautier (1997, 1999). I mainly restrict the data to the larger mammals (all ungulates), since the human use of carnivores, rodents, insectivores and lagomorphs (and even their contemporaneity) is at best problematic at the sites in question. The issue here, after all, is to highlight evidence of and variability in human adaptive responses, notably in terms of subsistence activities. Separate mention will be made of the use of fish, molluscs and birds, since their remains cannot be quantified in the same comparative fashion together with those of the ungulates. Microfaunal (and paleobotanical) data will at times be brought to bear as independent sources of evidence of environmental change.

La Riera Cave

This classic site (part of the Llera cluster of Upper Paleolithic habitation and cave art sites, that notably includes the adjacent locus of Cueto de la Mina) is located on the narrow coastal plan of eastern Asturias, midway (c.1.5 km north and south respectively) between the present shore of the Cantabrian Sea (Bay of Biscay) and the Sierra de Cuera coastal mountain range (maximum elevations: 715-1315m). La Riera's geographic coordinates are 43° 25'30" N x 4° 51'25" W x 30 m a.s.l. The early Tardiglacial shore was about 10 km distant, but half that distance by Preborealtimes. The 2000-2500m summits of the Picos de Europa lie only 20 km south of the cave. Re-excavation of the cave (discovered and first excavated by the Conde de la Vega del Sella in 1917 -18 [Vega del Sella 1930]) by G.A.Clark and myself in 1976-79 revealed some 30 thin natural levels spanning the period from just before the Solutrean through the Asturian Mesolithic.

Table 3.1. Ungulate faunas (NISP/MNI) from La Riera Levels 20-30 (from Altuna 1986)

Level	Cervus	Capreolus	Rangifer	Capra ibex	Rupicapra	Bovini	Sus scrofa	Equus
19/20*	929/?	2/1		209/?	1/1			1/1
20	706/13			163/6	1/1			1/1
21-23	983/16	6/3	2/2	156/4			2/2	17/3
24	431/11	23/3	5/2	350/11	8/2			8/2
25	37/4	6/1		7/1				
26	407/13	43/6		130/5	13/2		3/1	2/2
27	1088/28	74/6		178/7	52/6	10/2	28/4	11/3
28	86/8	18/1		4/1	1/1	11/2	3/1	
29	113/5	8/1		12/1	1/1			
30	39/?			1/1				

* combined in excavation

Table 3.2. La Riera relative frequencies of Cervus and Capra (based on NISP)

Level:	19/20	20	21-23	24	25	26	27low	27up	28	29	30
Cervus	81.3	81.1	84.3	52.2	74.0	68.1	69.6	83.8	70.0	84.3	97.5
Capra	18.3	18.7	13.4	42.4	14.0	21.7	15.6	7.9	3.3	9.0	2.5

Our 28 radiocarbon dates (despite some stratigraphic inversions and a few patently erroneous determinations, which are ignored here) range in age from 21 kya to 6.5 kya. These dates have recently been supported by 9 corrected radiocarbon determinations obtained by A. Craighead on shells from our levels (M.Gonzalez Morales, pers. comm.). The levels relevant to this topic are 20 (12,360±670 BP uncal.), 21-23 (12,800±110; 12,620±300), 24 (10,890±430; 11,480±75), 25, 26 (11,450±85), 27 lower (12,270±400; 10,990±70), 27 upper (10,630±120), 28 top (8830±90), 29 (8650±300), 29 top (6500±200). Levels 20-26 are attributable to the Upper Magdalenian, 27 to either the Terminal Magdalenian or Azilian, 28 to the Azilian and 29 to the Asturian. During this time the arboreal pollen percentage (AP) went from 5% in Level 20 (late Dryas I?) to 55% at the base of Level 30, which is a capping flowstone layer (Leroi-Gourhan 1986). The increase in AP is fairly gradual, but steady until Level 26, when it takes an abrupt jump up to 40% (Allerød?), only to fall back to 11% at the top of Level 27 (Dryas III?), then rising again in Level 28 and especially 29 (Preboreal and Boreal). There may be an erosive hiatus corresponding to at least part of Bölling (Laville 1986) and the presence of Dryas II here (as elsewhere in the Franco Cantabrian region) is problematic, although possibly corresponding to a slight dip in AP in Level 24.

The ungulate faunas from the P-HT levels at La Riera are listed in Table 3.1. Percentages of red deer and ibex based on NISP (numbers of identifiable specimens) are given in Table 3.2. The fundamental observation to be made about these assemblages that straddle the P-HT is that they change very little. With the exception of trace occurences of reindeer in Levels 22 and 24 (possibly corresponding to terminal Dryas I and Dryas II respectively ?), there are no cold macrofauna in the upper part of the La Riera sequence. Indeed, with the exceptions of Level 24 and a series of levels (1-6) at the base of the stratigraphy which show evidence of specialized ibex hunting occupations of the site, the percentage of red deer NISP is always between 70-85% of the ungulate total. There is of course a question as to whether the relatively high percentages of ibex in Levels 24 and 1-6 are the result of environmental conditions or human hunting choice. The high (42.4%) of ibex remains in Level 24 does correspond with an appearance of reindeer (a minimum of two individuals), yet reindeer is not present in Levels 1-6. Indeed, the levels that one would expect to be the coldest (i.e., those corresponding to the Last Glacial Maximum, c. Levels 7-12, the first of which does contain the Nordic root vole, Microtus oeconomus) have among the highest percentages of red deer remains of any of the La Riera levels. My preference is to see red deer as the dominant game species on plains and in valley bottoms under a very wide variety of climates and vegetation types in the Cantabrian region (see Straus 1981), while ibex (whose preferred steep, rocky habitat exists within 1-2 hours walk of La Riera in the coastal mountain range) could also be hunted at any time in this region, even very close to sea level, such that changes in the ratio of one to the other were probably mostly the result of hunting decisions and the nature and/or seaonality of site occupations. The only other possible mammalian indicator of cold climatic conditions is the reappearance in the cave of Microtus oeconomus (on its own or because of fox or raptor - not human - activity). On the other hand, boar and roe deer, both good indicators of woodlands are either singly or both present in Levels 23, 26, 27 and 28. Hence the evidence from Level 27 is somewhat contradictory, suggesting that the Dryas III cold snap did not succeed in totally eliminating woods in the rather southerly latitude of Asturias, as is indeed shown by the still relatively high AP in at least lower Level 27 (c.23-20%).

Bird bones were recovered from La Riera Levels 24-27 and were studied by A. Eastham (1986). Of significance is a ptarmigan (Lagopus mutus) in Level 24. This is probably another cold indicator, as the ptarmigan today lives at altitudes between 700-2000 m above sea level (including the Picos de Europa). It may have been forced down to the coastal plain during stadials, especially in wintertime. Otherwise this range of levels yielded very similar spectra of birds, including several ducks, geese, and smaller quantities of owls, eagles, kestrels, buntings, jays, thrushes, tits, etc.

Fish remains are present throughout virtually the whole La Riera sequence, from the early Solutrean on (Menéndez de la Hoz et al. 1986). However marine fish make their appearance only beginning in Level 24 times (sparids), replacing the salmon that were present throughout the Solutrean and complementing the salmon trout and brown trout that are present in the Solutrean and Magdalenian levels. Unidentified fish are particularly abundant in the Asturian *conchero* (Level 29). Whether the appearance of marine fish late in the sequence was simply a consequence of rising sea level or (also?) the result of human subsistence intensification (our preferred explanation, since the coast was never very far away), is difficult to ascertain. But fishing was always a part of the subsistence repertoire at La Riera throughout both the Late Upper Paleolithic and Mesolithic periods.

Similarly, the marine molluscan record, studied by J.A.Ortea (1986) shows constant shellfishing activity from early Solutrean times onward, but with diversification in the species gathered beginning in Upper Magdalenian times (i.e., from Level 19/20 onward). The appearance of the topshell (*Monodonta lineata*) beginning late in the sequence may be the result of warming waters (it ultimately replaces the periwinkle, *Littorina littorea*, especially in the Asturian shell–midden). But it is clear that there was a serious intensification of human shellfishing during Levels 27-28, both immediately before and after the 10 kya P-H boundary. Humans were exploiting not only the "easy to reach", estuarine molluscs (as before), but now also molluscs (plus urchins and crabs) that live in the open, wave-beaten littoral. This was probably caused by at least seasonal food stress among a growing, spatially constrained ("packed") human population. Limited oxygen isotope evidence on Asturian topshells and limpets suggests winter and autumn collection, but not summer or spring gathering (Deith and Shackleton 1986). This would also seem to support the subsistence stress hypothesis.

Dufaure Rockshelter
First excavated in 1900 by the Abbé Henri Breuil and Paul Dubalen, this site is part of a cluster of Tardiglacial loci at the base of the 30 m high, SSW facing Pastou Cliff on the boundary of Les Landes and the French Basque provinces in Pyrénées-Atlantiques, some 60 km north of the 2000 -2500 m high Pyrenean crestline. Dufaure' s geographic coordinates are 43° 30' N x 1°4' W x 35 m a.s.l.. The Pastou sites are about 40 km from the present Bay of Biscay shore. Because of the Capbreton undersea canyon which extends almost to the present shore 15 km north of the mouth of the Adour, this minimal distance would not have changed under glacial conditions, but the rest of the seashore would have been 12-20 km further west than it is today. Dufaure and the other Magdalenian (and Azilian) sites (Duruthy, Grand Pastou and Petit Pastou) are directly above

the edge of the "Würm II" alluvial terrace of the Gave d'Oloron, 6 km upstream of its confluence with the Gave de Pau. These two Pyrenean rivers in turn join the Adour another 6 km downstream. The sites are directly in front of a major ford formed by an ophite outcrop across the Oloron. The huge site of Duruthy, with a stratigraphic sequence similar to that of Dufaure (Arambourou 1978), lies 230 m to the west.

Our excavation at Dufaure (on the terrace and talus in front of the small rock shelter) in 1980-84 yielded a stratigraphy including early-middle, middle, upper and final Magdalenian and Azilian levels. The 14 radiocarbon dates span the period between 14,600-9600 BP. Sedimentological and palynological analyses (Laville 1995; Marguerie 1995) indicate the presence of late Dryas I, (early Bölling), Allerød, Dryas III and Preboreal. The existence of a Dryas II episode is problematic, as it is in deep sea cores from the Bay of Biscay. There seems to have been a significant episode of erosion in late Bölling . The radiocarbon determinations are for Levels 6 base (14,640± 230; 14,590± 100 BP uncal.), 6 top (14,020± 340), 5 base (14,570± 390), 5 (12,990± 270; 12,690± 230 BP), 4 base (12,260± 400), 4 (12,030± 280; 11,750± 300), 4 top (10,910± 220), 3 lower middle (10,310± 270), 3 middle (9810± 100, 9750± 110, 9600± 290). Despite the attribution to Allerød and the 12.3-11 kya dates, Level 4 is assignable to the Upper/Final Magdalenian, rather than to the Azilian (which already "existed" elsewhere in France). Three cylindrical section antler harpoons (2 bases and 1 whole unilaterally barbed harpoon) were found near the base of Level 4. In order to study changes in ungulate faunal assemblage composition throughout the course of the critical Allerød period, Level 4 was divided into lower, lower middle, upper middle and upper portions, based on groups of excavations spits, which, in turn, corresponded to prehistoric, man-made cobblestone pavement units. The Dufaure ungulate faunal assemblages are presented in Table 3.3.

Unlike the La Riera sequence across the P-HT, Dufaure demonstrates some clear shifts in the ungulate game that was hunted between late Dryas I and Preboreal. Saiga antelope only appears (as elsewhere in Aquitaine [Delpech 1992]) during the Middle Magdalenian. Bison is gradually replaced by aurochs; boar and especially roe deer steadily increase, while horse fairly steadily decreases (from 27% of NISP in Levels 6+5 to <1% in Level 3). The most remarkable aspects of the archeofaunas from Dufaure are:

1.) the "boom" in reindeer hunting in the earlier Level 4 Upper Magdalenian occupations, under at first cooler, more open conditions of "Dryas II", and then under relatively temperate, parkland conditions of Allerød, and
2.) the progressive replacement of reindeer by red deer throughout the final Magdalenian and Azilian (Table 3.4).

Indeed, one of the interesting results of the research at Dufaure was to confirm the late presence of reindeer in this peri-Pyrenean area, as earlier (and controversially) argued by Arambourou (1978) and Delpech (1978) at Duruthy. A possibly isolated Rangifer herd managed to survive and was hunted by humans throughout Allerød and as recently as Dryas III and Preboreal along (and presumably in) the

23

Table 3.3. Ungulate faunas from Dufaure Levels 6-3 (from Altuna and Mariezkurrena 1995)

Level	Cervus	Capreolus	Rangifer	Saiga	Bovini	Bison	Bos primigenius	Sus scrofa	Equus
6	5/1		2/1	1/1					4/1
5	39/2	3/1	253/7		94/3	3/1		1/1	150/5
4 low	82/3		382/7		28/4	5/4			42/3
4 l-m	166/4	1/1	526/8		61/2	2/1		1/1	46/2
4 u-m	303/6	22/2	965/13		108/5	8/?	9/?	6/1	100/4
4 up	421/9	32/3	483/9		100/6	5/?	5/?	6/1	92/4
3	673/8	67/3	25/4		15/1			10/2	7/1

Table 3.4. Dufaure relative frequencies of *Cervus* and *Rangifer* (based on NISP)

Level:	6+5	4 low	4 l-m	4 u-m	4 up	3 low	3 up
Cervus	7.7	15.0	20.3	19.7	36.5	85.8	80.5
Rangifer	44.6	70.0	64.4	62.7	41.8	3.4	2.4

western Pyrenees, long after this cold adapted species had become extirpated in most other (lowland) regions of southern France such as the Périgord (Delpech 1983), and even on the plains of northern France (Fagnart 1997), from all of which areas reindeer had disappeared before the beginning of Allerød (Fagnart 1997). However, there is evidence that the Pyrenees was not the only mountain chain in France to harbor late surviving reindeer herds; others include the Massif Central (e.g., Surmely 1998; Bracco 1991), Jura (e.g., David 1992; David and Richard 1989) and French Alps (e.g., Binz and Desbrosse 1979; see also Delpech 1989). The Dufaure and Duruthy data clearly show a phenomenon of displacement in action, as red deer (expanding out of Spain) took the place of reindeer. The latter survived in dwindling numbers despite dramatic habitat changes (notably increased woodlands, with AP rising overall from about 17 to about 61% from Level 5 to Level 3 top [Marguerie 1995]). They presumably managed to do so by being able to ascend to high summer pastures in the Pyrenees. In fact, seasonality indicators (tooth eruption and/or cementum, antlers, migratory birds, fish, mole ethological evidence) all point to cold season human occupations (and hunting) at Dufaure, Duruthy and Grand Pastou. On the other hand, both Dufaure and Duruthy data show that reindeer hunting was not overwhelmingly important in the Middle Magdalenian; it really took off only with the Upper Magdalenian, perhaps as a result of the development of very specialized hunting tactics and weapons, and of a particular overall subsistence and mobility strategy on the part of Allerød human bands in this "reindeer refugium". At Duruthy, reindeer NISP rose to 72% of the ungulates in the combined pavement layers (Level 3) of the 11,200 BP Final Magdalenian. It had earlier represented only 7.5-25% of the ungulates, and fell back to 24% of the small combined Azilian assemblage. As at Dufaure, the numbers of reindeer remains decreased with time within the Azilian horizon itself (Delpech 1978), but Rangifer managed to hang on (presumably now acting more like the "woods caribou", i.e., more solitary smaller bands and less migratory [Spiess 1979]) well into Preboreal, until the final abandonment of Duruthy by foraging peoples.

Thus, both biogeographic factors involving shifts in red deer and reindeer vicarage and (ultimately short-sighted) human subsistence choices (i.e., to increasingly specialize in reindeer hunting during the cold season in the Pastou area) seem to have been involved in the history of faunas at these sites.

Other interesting aspects of the Dufaure faunas include the presence of arctic fox only in Level 4 (Upper Magdalenian) and the virtual absence of cold loving *Microtus ratticeps/oeconomus* in Azilian Level 3 (and its complete absence in the Duruthy Azilian), after relative abundance in the Magdalenian levels (Eastham 1995a). The avifaunas, although having some continuity in represented habitats, show the disappearance of snowy owl, of willow grouse, and of "winter guest" species (all found in the Magdalenian levels) in the Azilian of Dufaure. Some of the changes in the bird spectra between the Magdalenian and Azilian horizons are also indicative of increasing forestation, despite the continued existence of open grasslands on the plateau above the site at least in Dryas III (Eastham 1995b). There is no reported evidence of fishing in the Azilian at either Dufaure or Duruthy. At Duruthy, Final Magdalenian fishing of salmon was significant, followed by that of sea trout and pike; harpoons are common (LeGall and Martin 1996). Fishing occurred mainly in fall and winter. In contrast, despite fine screening, fish are virtually absent even in the Final Magdalenian of Dufaure (1 trout and 1 pike - the latter taken in spring) and harpoons are very rare.

Despite the facts that La Riera and Dufaure are at virtually the same latitude and are only 300 km apart, there are striking differences between them (and their respective regions: Vasco-Cantabria and Gascony) in terms of the histories of their archeofaunal assemblages across the P-HT. There was essential continuity in animal communities and increasing human subsistence diversification in the former versus significant change in animal communities (albeit temporally retarded due to the "mountain chain effect" vis à vis central and northern Aquitaine) and increased human subsistence intensification followed by a "crash" in the

24

latter. The Magdalenian and Azilian foragers in the northern shadows of the Canabrian Cordillera and of the Pyrenees lived in very different worlds, geographically and ecologically, dependant on significantly different resource structures on a narrow coastal strip and on the southern edge of the vast plains of Aquitaine respectively. These differences are clearly reflected in their subsistence evidence, including not only ungulates, but also other resources from shellfish to birds. The P-HT was clearly a more abrupt, sharp, significant "event" in SW France than in NW Spain - faunistically and in terms of human adaptations. It may have been delayed somewhat (vis à vis the Périgord, for example) by the late survival of reindeer in and along the Pyrenees, but even there, the "break" did eventually come. And it was very disruptive of human adaptations that had for several millennia come to depend significantly upon that gregarious, docile, highly mobile cervid.

Bois Laiterie Cave and Pape Rockshelter

It was not long before the Pleistocene Holocene boundary (i.e., c.13 kya) that humans reoccupied northwestern Europe (northern France, England, Benelux and Germany) after the period of abandonment that had coincided with the Last Glacial Maximum. As soon as conditions (especially humidity) improved enough for there to be vegetation adequate to support a large mammalian biomass, humans began again to expand their range northward late in Dryas II and especially in Bölling. This was a time of considerable ecological contrast and flux, especially in the higher latitudes. In order to sample the archeofaunal record across the P-HT at the northern frontier of human settlement, I combine evidence from two nearby sites that I recently excavated in association with M. Otte and his team, in the deeply entrenched upper Meuse valley on the western edge of the Ardennes uplands of Namur Province in Wallonia (southern Belgium): Bois Laiterie Cave (Upper Magdalenian and early Mesolithic, incorrectly sometimes known as "Burnot") and Pape Rockshelter (middle Mesolithic).

Bois Laiterie (BL) is a small, drafty, north-facing, double-mouthed cave that dominates a strategic western side gorge of the Meuse canyon immediately upstream of their confluence. The site is close to a variety of strongly differentiated habitats: the Meuse valley bottom and slopes, the talweg leading up to the 250 m high interfluve between the Meuse and Sambre, north-, southeast- and west-facing slopes, and the plateau dominating the rivers. These differences would have been accentuated under Tardiglacial conditions. Geographical coordinates of the site are 50° 21' N x 4° 52' E x 120 m a.s.l. BL is c. 35 m above the valley floor on a very steep, rocky slope.

Neolithic and Mesolithic deposits that capped the BL deposit had been largely removed by clandestine looters before we undertook our complete excavation of the underlying Magdalenian horizon (levels YSS+BSC) in 1994-95. This horizon (separated from the Mesolithic component by sterile sands) had been discovered in 1991 through limited testing by Ph.Lacroix and dated to 12,660± 140 BP by an AMS determination on an antler sagaie (Charles 1996). We obtained two other AMS dates for Level YSS, top and base respectively: 12,625± 117 and 12,665± 96 BP. These three statistically identical dates, together with evidence of refits,

definite artifact, manuport and faunal concentrations suggestive of distinct activity areas, the "fresh", generally undamaged condition of the artifacts and bones, and micromorphological analyses indicating only limited movement or mixing, all point to a relatively undisturbed Magdalenian horizon with considerable stratigraphic integrity (Straus and Martinez 1997; Courty 1997). Although perhaps resultant from several short, closely spaced human visits to this uncomfortable, but strategically located cave, the horizon is essentially that of a single cultural component and was treated as such in analysis. Along with several other caves in southern Belgium, BL testifies to a Magdalenian occupation during the traditional Bölling pollen zone (sensu lato). The presence of trees and shrubs including some mesophile taxa within the context of a mosaic landscape (still including open areas) is testified to by pollen and paleobotanical analyses at BL (Emery Barbier 1997; Pernaud 1997) and other contemporary sites in the region. The mosaic nature of the vegetation around the cave mouth is also hinted at by the malacofaunas (López-Bayón et al. 1997).

The remnant BL early Mesolithic component consists of a breccia adhering to the eastern and southern cave wall. It was "discovered" when we submitted an exposed human footbone we had extracted for AMS dating: 9235± 85 BP. Subsequently, I.López-Bayón et al. removed a large block of the breccia and extracted more human remains. The total human MNI count is now at least 6 and associated mammalian faunal remains were also found, but there are no Mesolithic artifacts (Otte and Straus 1997; Toussaint et al. 1998). BL thus joins a growing list of 6 other Mesolithic human burial caves in the Belgian upper Meuse and Sambre valleys, all of which have very few or no associated artifacts and all of which date between 9.0-9.6 kya (uncal.).

Not only are Dryas II, Allerød and Dryas III deposits absent from the sites that we have excavated in Belgium, but also archeological levels (and hence archeofaunas) of these periods from well excavated (or any) sites are very scarce in Belgium overall. In fact, the only exceptions are the Ahrensburgian or Tjongerian levels in Remouchamps and Coléoptère caves in eastern Belgium, which are respectively dated or attributed to Dryas III. Reindeer (together with steppe lemming) makes a last appearance in Belgium at this time (Cordy 1991).

The P-HT chronostratigraphic sequence for southern Belgium is continued by the later Mesolithic levels in Pape Rockshelter, 15 km upstream along the Meuse from BL, which we excavated in 1993-94. Pape is a small rockshelter at the foot of the 100 m high Freyr Cliff on the right (east) side of the river. The top of the talus is 8.5 m above the present (artificially maintained) level of the Meuse. The shelter faces southwest. Its geographic coordinates are 50° 13' N x 4° 53' E x 100 m a.s.l. The lowest Mesolithic level (22+22.1) dates by AMS to 8780± 85 and 8756± 83 BP; the middle level (21) dates to a statistically identical 8717± 85 BP. The uppermost Mesolithic level, however, dates to about 1000 years later: 7843± 85 BP. In short, this sequence spans the traditional Boreal pollen zone, with temperate, humid, wooded environmental conditions (as is shown by paleobotanical, malacological and

Table 3.5. Ungulate faunas from Bois Laiterie and Pape (from Gautier 1997, 1999; López Bayón et al. 1996)

Level	Cervus	Capre-olus	Alces	Rangi-fer	Capra	Rupi-capra	Bison	Bos primi-genius	Ovibos	Sus scrofa	Equus caballus	Equus hemi-onus
BSC			2/1	4/1	4/1	1/1	1/1				8/2	
YSS	3/2		3/1	40/3	24/1	10/1	2/1		11/1		46/2	1/1
Breccia	x									x		
22*	5	1							2	11		
21*	7	4								2		
20*	18	14							5	25		

x = present; *NISP only at Pape

micromammalian studies at Pape [Pernaud 1999; López-Bayón et al. 1999; Gautier 1999]).

The ungulate archeofaunal assemblages from BL levels BSC, YSS and Breccia, and from Pape levels 22, 21 and 20 are summarized in Table 3.5. The Magdalenian, Bölling age faunal assemblage from BL is extraordinary in its diversity and apparent incongruity, combining as it does cold, open steppe-tundra taxa (reindeer, bison, muskox, horse) *and* the marshy, woodland moose. The co-existence of reindeer with traces of red deer is certainly suggestive of environments in flux, although both cervids are certainly capable of both grazing and browsing and of living in both open vegetation habitats and woodlands as they do today in different regions of the northern hemisphere. The ecological significance of the ass is uncertain, but it is likely to have favored dry steppes, like the saiga antelope (not represented at BL, but present in the radiometrically contemporary Magdalenian of nearby Chaleux Cave [Otte 1994], where it is associated with ass, horse, muskox, reindeer, red deer, bison and/or aurochs, chamois and ibex) (Gautier 1997; Patou-Mathis 1994). Similarly unusual faunas have been found not only at Chaleux, but also at Goyet, also in Namur Province: reindeer, red deer, roe deer, bison and/or aurochs, muskox, ibex and chamois (Germonpré 1997). The presence of roe deer is particularly indicative of the existence of woodlands, where as the caprines really only signify the obvious presence of abrupt, rocky slopes and cliffs, although chamois can be a true woodland dweller.

There are two possible explanations for these peculiar faunas from these three Belgian Magdalenian sites: mechanical mixture or the real co-existence of "disharmonious" faunas during the Bölling in this northerly region (cf. FAUNMAP 1996). The mixing hypothesis is certainly plausible in the cases of the 1860s excavations by Edouard Dupont in Goyet and Chaleux, although careful work has been done to separate out clear or likely Holocene intrusives (e.g., sheep, boar) (Germonpré 1997; Charles 1996). However, in the modern excavations at Chaleux and Bois Laiterie these taxa are really physically associated, still leaving, of course, the possibility of "telescoping" and the creation of palimpsests that would lump Dryas I and Bölling faunas (Gautier 1997). This seems rather implausible to me, due to the intimate contact and good physical condition of the remains of the supposedly discordant animals in relatively thin levels. It should be noted that one musk ox bone each has been dated from Chaleux (this one a phalanx with cut marks) and

Goyet: 12,860± 140 and 12,620±90 BP (uncal.) respectively. Another musk ox bone from the Magdalenian layer in Trou Da Somme has recently yielded an AMS date nearly identical to the date from nearby Chaleux (R.Miller, personal communication). These dates fall within the traditional age range of the Bölling, although some would argue that in reality they are still within late Dryas I (e.g., Germonpré 1997). Regardless of this rather semantic issue, the co-occurence of "cold" and more "temperate" faunas in several Belgian Magdalenian sites is the interesting fact. It should be noted that at BL, the YSS microfaunas (rodents, lagomorphs and insectivores) display striking combinations of temperate (open and wooded), arctic, and continental (dry and humid) taxa (Cordy and Lacroix 1997). The same general range of seemingly "discordant" microfauna is found in the Magdalenian horizon at Chaleux (Cordy 1994), as well as at Presles I in the Sambre valley and at Walou in eastern Belgium (Cordy 1991). Steppe lemming appeared to decline in favor of various woodland taxa, but it and other cold, dry and/or open environment loving species did continue to co-exist with the "newcomers". Both common and arctic fox are found together in the Magdalenian assemblages from BL and Chaleux (Gautier 1997; Patou-Mathis 1991). Although there may have been minor climatic fluctuations within the c.800 years of Bölling, this does not seem to be the explanation for the phenomenon of disharmonious faunas described here.

Other human subsistence resources at BL were fish (brown trout, burbot and grayling) and maybe some of the birds (Van Neer 1997; Deville and Gautier 1997). The birds include indicators of both open (notably willow grouse) and wooded habitats as well as of aquatic ones in Magdalenian times. This evidence reinforces the Bölling environmental mosaic hypothesis.

In line with the the microfaunal and admittedly limited pollen and paleobotanical data, I suggest that this human reoccupation of the Ardennes fringes, at >50° N latitude happened right at the cusp of a major ecological change, from open, arctic like environments to more temperate, wooded ones. During that brief moment, given the marked relief of Wallonia, there continued to exist habitats adequate for the likes of reindeer, horse, ass and muskox (e.g., the exposed plateaux and ridges), while new, more wooded habitats were being created (e.g., in the canyon-like valleys - especially on sheltered southerly exposed slopes and well watered bottomlands). Sites such as BL, Chaleux and Goyet were immediately adjacent to all such habitats, so that their hunters could easily "sample" animals from radically different settings.

After the hiatus in our archeofaunal record for Dryas II-III (save for the final presence of reindeer in Dryas III, as noted above), our indications for Preboreal are meagre, including not only red deer, but also boar in the small early Mesolithic breccia sample from BL. The wooded environments are further confirmed by the Boreal assemblages from Pape, dominated by boar, roe deer and red deer. The limited presence of aurochs, however, hints at the existence of some clearings. The microfaunal assemblages from Pape also point to generally wooded conditions - and increasingly so through time (Gautier 1999). The malacological analysis points to predominantly wooded habitats, but with the existence of clearings, and a mixture of xerophile, mesophile and moisture loving molluscs (López-Bayón et al. 1999). The malacofauna and birds (Deville 1999) suggest the existence of colder, more open environments in the non cultural, pre-Mesolithic strata (23-24) (Preboreal and Dryas III?). The avifaunas of the Mesolithic levels are rich and diverse, including many aquatic, woodland and cliff dwelling species. However, many or most may have been the result of mammalian carnivore and/or raptor activity or of natural deaths. There is no cut mark evidence for human agency, although this is possible especially in the cases of some of the larger waterfowl.

The Pape Mesolithic levels are (not surprisingly) quite rich in fish remains, with no significant differences between the 9 kya and 8 kya horizons (Van Neer 1999). About 77% of the 212 identifiable fish remains from levels 20 and combined 21+22 (with lenses) belong to the carp family (several species). There are also eel, shad, catfish, pike, salmonids and perch. In addition, there are numerous unidentified fish remains, all testifying to the wealth of the aquatic resources available to and exploited by Mesolithic foragers, in contrast to the lesser diversity of fish represented in the Magdalenian at BL and other Belgian sites.

In short, the Belgian evidence suggests that the northward movement of humans in the late Tardiglacial was part of or linked to a wider ecogeographical phenomenon: the re-expansion of vegetation and animals out of southerly refugia of western Europe into the rapidly changing environments above 48° N latitude. As soon as humidity and temperatures had risen sufficently, these environments were once again capable of supporting a biomass adequate to feed hunter-gatherer bands with a wide variety of ungulate taxa, some (notably herd species) held over from the arctic-like conditions of Dryas I and others newly arrived from further south. It was the highly contrasted relief (despite relatively low elevations) of the Ardennes uplands and canyons that made such an ecological mosaic possible, even if for only a relatively short time before the definitive extirpation of reindeer, saiga, muskox, ass and bison. This was a world without modern analogues. In contrast, the world of Mesolithic foragers in the same region was a more familiar one of woodlands (albeit denser and more dominated by mixed deciduous, mast-rich taxa in Boreal than in Preboreal times), populated by less gregarious boar, roe and red deer, but whose rain-swollen rivers were richer in fish and fowl.

Conclusions

The three examples presented here amount to a controlled comparison of differences in regional ungulate (and other) faunas and in their exploitation by humans. We have focused on the period between Bölling and Boreal insofar as possible, using carefully excavated and fully analyzed materials from radiocarbon dated stratigraphic contexts. Within the space of the only 300 west-east kilometers between La Riera in northern Spain and Dufaure in southwesternmost France and at the same latitude, there were striking differences in the nature of the P-HT: fundamental continuity in resources and diversified subsistence strategies in the coastal zone of Asturias versus extirpation of the key game species (reindeer) and an abrupt failure of the relatively specialized system dependent on its hunting in the western French Pyrenean piedmont. The "invasion" of reindeer territory in southwestern France by red deer from Iberia took place throughout the late Tardiglacial. This cervid replacement was somewhat retarded in the Pyrenean region, but nonetheless inevitable, despite centuries of coexistence of the two taxa in the vicinity of the Pastou sites especially during Allerød. An even more dramatic case of "disharmonious" fauna is that presented by the Bölling era Magdalenian occupation of Belgium, seven degrees of latitude and 900 km to the north. Human recolonization of this region was made possible by the environmental changes of the Tardiglacial and in particular by the unusual convergence of cold, dry, open habitat-loving *and* more temperate, humid, woodland-loving game animals during Bölling, a brief time that, at this latitude, straddled glacial and interglacial conditions. This cohabitation of taxa, so advantageous to human hunters, was made possible under the mosaic circumstances of vegetation, water access and exposure afforded by the relief of the Ardennes and its fringes. There was a brief "boom" in human settlement in Belgium during this specific, ecologically felicitous time - followed by at least partial abandonment of the territory in Dryas II and only very sparse reoccupation in Allerød and Dryas III. The north was a land of radical ecological changes and of discontinuous human settlement When humans once again came to be fairly abundant on the Belgian landscape in Preboreal and Boreal, it was a totally different landscape, with many different resources and new ways of exploiting them - notably broad-spectrum subsistence diversification, including the exploitation of many smaller, more solitary, or aquatic and avian species. The change was major and, relative to the south, relatively late.

Acknowledgements

I wish to thank the colleagues with whom I have worked in the excavations of La Riera, Dufaure, Bois Laiterie and Pape, notably G.A.Clark, M.González Morales, the late R.Arambourou, M.Otte and Ph.Lacroix, and especially the archeozoologists on these projects, notably J.Altuna, K. Mariezkurrena, A.Eastham, A.Gautier and I.López Bayón, none of whom, however, are responsible for my potential misinterpretations of their findings. My work in Spain, France and Belgium has been supported by numerous grants from the National Science Foundation, the National Geographic Society, the L.S.B.Leakey Foundation, and the University of New Mexico, as well as by grants to M. Otte from the Regional Government of Wallonia.

References

Altuna, J., 1986. The mammalian faunas from the prehistoric site of La Riera. In Straus and Clark, ed., 1986, pp. 237-274, 421-480.

Altuna, J. and K.Mariezkurrena, 1995. Les restes osseux de macromammifères. In Straus, ed., 1995c pp.181-212.

Arambourou, R., ed., 1978. *Le Gisement Préhistorique de Duruthy à Sorde l'Abbaye* Société Préhistorique Francaise. Paris: Mémoires 5.

Bintz, P. and R. Desbrosse, 1979. La fin des temps glaciaires dans les Alpes du Nord et le Jura méridional. In *La Fin des Temps Glaciaires en Europe*, ed. D.de Sonneville Bordes. Paris: CNRS, pp.329 –255.

Bracco, J-P., 1991. Le Paléolithique supérieur du Velay. *Bulletin de la Société Préhistorique Francaise* 88, 114-121.

Charles, R., 1996. Back into the North: the radiocarbon evidence for the human recolonisation of the north-western Ardennes after the Last Glacial Maximum. *Proceedings of the Prehistoric Society* 62, 1-17.

Cordy, J M., 1991. Palaeoecology of the Late Glacial and early Postglacial of Belgium and neighbouring areas. In *The Late Glacial in North West Europe*, ed. N.Barton, A.Roberts and D.Roe. London: Council for British Archaeology Research Report 77, pp.40-47.

Cordy, J M., 1994. Analyse paléoécologiques des micromamifères tardiglaciaires de la Grotte de Chaleux. In Otte, ed., 1994, pp.178-192.

Cordy, J M. and P.Lacroix, 1997. Bio- et chronostratigraphie de la Grotte du Bois Laiterie à partir des microvertébres. In Otte and Straus, eds., pp.161-176.

Courty, M A., 1997. Etude micro stratigraphique de la Grotte du Bois Laiterie. In Otte and Straus, eds., pp.113-140.

David, S., 1992. Le peuplement magdalénien dans le nord-est de la France. In *Le Peuplement Magdalénien*, ed. J P.Rigaud. Paris: CTHS, pp. 87-96.

David, S. and H.Richard, 1989. Les cultures du Tardiglaciaire dans le nord est de la France. In *Le Magdalénien en Europe*, ed. J P.Rigaud. Liège: ERAUL 38, pp.101-154.

Deith, M. and N.Shackleton, 1986. Seasonal exploitation of marine molluscs: oxygen isotope analysis of shell from La Riera Cave. In Straus and Clark, eds., pp.299-314.

Delpech, F., 1978. Les faunes magdaléniennes et aziliennes du gisement de Duruthy. In Arambourou, ed., pp.110-116.

Delpech, F., 1983. *Les Faunes du Paléolithique Supérieur dans le Sud-Ouest de la France* . Paris: CNRS.

Delpech, F., 1989, L' environnement animal des magdaléniens. In *Le Magdalénien en Europe*, ed. J P.Rigaud. Liège: ERAUL 38, pp.5-30

Delpech, F., 1992. Le monde magdalénien d'après le milieu animal. In *Le Peuplement Magdalénien*, ed. J P. Rigaud, H.Laville and B.Vandermeersch. Paris: CTHS, pp.127-135.

Deville, J. and A.Gautier, 1997. The avifauna of la Grotte du Bois Laiterie. In Otte and Straus, eds., pp.215-218.

Eastham, A., 1986. The La Riera avifaunas. In Straus and Clark, ed., pp.275-284.

Eastham, A.1995a. La microfaune. In Straus, ed., pp.235-245.

Eastham, A., 1995b. L'écologie avienne. In Straus, ed., pp. 219-233.

Emery Barbier, A., 1997. Analyse palynologique de la Grotte du Bois Laiterie. In Otte and Straus, ed., pp.141-142.

Eriksen, B. and L.Straus, ed., 1998. As the World Warmed: Human Adaptations across the Pleistocene Holocene Boundary. *Quaternary International* 49/50. Oxford: Elsevier Science.

Fagnart, J-P., 1997. *La Fin des Temps Glaciaires dans le Nord de la France*. Paris: Société Préhistorique Francaise, Mémoire 24.

FAUNMAP, 1996. Spatial response of mammals to Late Quaternary environmental fluctuations. *Science* 272, 1601-1606.

Gautier, A., 1997. The macromammal remains of la Grotte du Bois Laiterie. In Otte and Straus, eds., pp.177-196.

Gautier, A., 1999. The mammalian remain of the Mesolithic and earlier strata in Abri du Pape. In Straus et al., eds.

Germonpré, M., 1997. The Magdalenian upper horizon of Goyet and the late Upper Paleolithic recolonisation of the Belgian Ardennes. *Bulletin de l' Institut Royal des Sciences Naturelles de Belgique* 67, 167-182.

Laville, H., 1986. Stratigraphy, sedimentology and chronology of the La Riera Cave deposits. In Straus and Clark,ed., pp.25-56.

Laville, H., 1995. Caractéristiques et signification des dépôts. In Straus, ed., pp.33-47.

LeGall, O. and H. Martin, 1996. Pêches et chasses aux limites Landes/Pyrénées. In *Pyrénées Préhistoriques*, ed. H.Delporte and J.Clottes. Paris: CTHS, pp.163-172.

Leroi Gourhan, Arl., 1986. The palynology of La Riera Cave. In Straus and Clark, eds., pp.59-64.

López Bayón, I., P.Lacroix and J-M.Léotard, 1997. Etude des restes malacologiques de la Grotte du Bois Laiterie. In Otte and Straus, eds., pp.145-160.

López Bayón, I., P.Lacroix and J-M.Léotard, 1999. Etude des restes malacologiques de l' Abri du Pape. In Straus, et al., ed.

López Bayón, I., L.Straus, M.Otte, et al., 1996. La Grotte du Bois Laiterie, du Magdalénien au Mésolithique: différences comportementales. *Notae Praehistoricae* 16, 63-74.

Marguerie, D., 1995. Etude palynologique. In Straus, ed., pp.49-53.

Menéndez de la Hoz, M., L.Straus and G.Clark. 1986. The icthyology of La Riera Cave. In Straus and Clark, ed., pp.285-288.

Ortea, J., 1986. The malacology of La Riera Cave. In Straus and Clark, eds., pp.289-314.

Otte, M., ed., 1994. *Le Magdalénien du Trou de Chaleux*. Liège: ERAUL 60.

Otte, M. and L. Straus, ed., 1997. *La Grotte du Bois Laiterie*. Liège: ERAUL 80.

Patou-Mathis, M., 1994. La grande faune. In Otte, ed., pp.172-178.

Pernaud, J-M., 1997. Le site du Bois Laiterie: Rapport de l' analyse anthracologique des niveaux du Magdalénien supérieur. In Otte and Straus, ed., pp.143-144.

Pernaud, J-M., 1999. Contribution de l'anthracologie à la connaissance du paléoenvironnement des occupations mésolithiques de l' Abri du Pape. In Straus et al., ed.

Spiess, A., 1979. *Reindeer and Caribou Hunters*. New York: Academic Press.

Straus, L., 1981. On the habitat and diet of *Cervus elaphus*. *Munibe* 33, 175-182.

Straus, L., 1991. The Epipaleolithic and Mesolithic of Cantabrian Spain and Pyrenean France. *Journal of World Prehistory* 5, 83-104.

Straus, L., 1992. To change or not to change: the Late and Postglacial in Southwest Europe. *Quaternaria Nova* 2, 161-186.

Straus, L., 1995a. A través de la frontera Pleistoceno-Holoceno en Aquitania y en la Península Ibérica. In *El Final del Paleolítico Cantábrico*, ed. A.Moure and C.González Sainz. Santander: Universidad de Cantabria, pp.341-363.

Straus, L., 1995b. Diversity in the face of adversity. In *Los Ultimos Cazadores*, ed. V.Villaverde. Alicante: Instituto de Cultura Juan Gil Albert, pp.9-22.

Straus, L., ed., 1995c. *Les Derniers Chasseurs de Rennes du Monde Pyrénéen: L' Abri Dufaure*. Paris: Société Préhistorique Francaise, Mémoires 22.

Straus, L. and G. Clark, ed., 1986. *La Riera Cave*. Anthropological Research Papers 36, Tempe.

Straus, L, B.Eriksen, J.Erlandson and D.Yesner, ed., 1996. *Humans at the End of the Ice Age*. Plenum Publishing Corp., New York.

Straus, L and A. Martínez, 1997. Site formation/disturbance processes, spatial distributions, site structure and activity areas. In Otte and Straus, eds., pp.65-112.

Straus, L, M.Otte, J M.Léotard and I.López Bayón, 1999. *L' Abri du Pape*. Liège: ERAUL.

Surmely, F, 1998. Découverte d' un important gisement de plein air du Magdalénien final: 'Le Pont de Longues'. *Bulletin de la Société Préhistorique Francaise* 95, 449-456.

Toussaint, M., I. López Bayón, M. Otte, L. Straus, et al., 1998. Les ossements humains du Mésolithique ancien de la Grotte du Bois Laiterie. In *Actes de la Sixième Journée d' Archéologie Namuroise*, ed. J. Plumier and C. Duhaut. Namur: Ministère de la Région Wallonne, pp.33-50.

Van Neer, W., 1997. Fish remains from the Upper Magdalenian in the Grotte du Bois Laiterie. In Otte and Straus, eds., pp.205-213.

Van Neer, W., 1999. Fish remains of the Abri du Pape. In Straus et al., ed.

Vega del Sella, Conde de la, 1930. *Las Cuevas de la Riera y Balmori*. Madrid: Comisión de Investigaciones Paleontológicas y Prehistóricas, Memoria 38.

MODELING OCCUPATION INTENSITY AND SMALL GAME USE IN THE LEVANT

Todd A. Surovell

Department of Anthropology, University of Arizona, Tucson, AZ 85721

Introduction

The Pleistocene-Holocene transition in the Levant is marked by dramatic shifts in human lifeways. With the termination of the Ice Age, we see the end of economic systems based on the collection of wild foods and the rise of settled farming communities. This much is clear. What remains to be determined is the nature and organization of the factors shaping and guiding this change. Certain prime movers have always taken the forefront: population pressure (Binford 1968; Cohen 1977; Flannery 1969), social competition (Bender 1978; Hayden 1990), and environmental change (Bar-Yosef and Meadow 1995; Childe 1928; Wright 1977) to name a few. Also common components of these discussions are intensification and/or resource broadening, which are generally believed to be omens of an imminent agricultural horizon. Unfortunately, these ideas remain largely conjectural as they suffer from few actual tests of their applicability to models of agricultural origins.

This paper will explore how Levantine small game assemblages can inform us about occupation intensity and human demography before, during, and after the Neolithic revolution. Using computer modeling of small game populations I develop a simple framework for exploring the effects of increased human demographic pressure on populations of three small game taxa utilized throughout the Levantine sequence since at least the Middle Paleolithic (Stiner et al. 1999, n. d.): *Testudo graeca*, the Mediterranean spur-thighed tortoise, *Lepus capensis*, the cape hare, and *Alectoris chukar*, the chukar partridge. Using this approach, Munro (this volume) then evaluates changes in occupation intensity through time at Hayonim Cave focusing especially on the Natufian deposits.

The basic premise of this paper is that human beings will have direct effects on the ecosystems they occupy. The more people there are on the landscape, and the greater the length of time those people remain in one place in an ecosystem, the greater the effects will be. Therefore, reduced human residential mobility, on a local scale, and population pressure, on a regional scale should be manifested in faunal assemblages as animal populations respond to rising pressures from greater exploitation. Certain populations will suffer more than others, and through time, faster reproducing species will increase in number in the archaeological record, while slower reproducers will decrease and possibly be driven to extinction. This dynamic then provides a means of estimating the intensity at which human are affecting ecosystems on a local and/or regional scale, which in turn may be able to tell us something about the distribution of people on the landscape in space and time, particularly when combined with other lines of evidence.

Small animals are ideal for exploring this phenomenon. Large animals, hampered by their body size, tend be rather slow reproducers (Bonner 1965; Pianka 1970; Charnov 1993). Reproduction is postponed to provide ample time for growth to adult size. Also, offspring are usually limited to only one or two individuals per year due to the costs of raising and feeding large-bodied young. Small animals, however, show a multitude of reproductive strategies because they are not limited by body size constraints. While we tend to equate small body size with high reproductive potential and maximization of population growth rates (e.g. Hayden, 1981), this is a very mammalian view of the world. If we take a moment to step outside of our warm bodies and consider other strategies, we see a very different picture. In fact, the only things that all small game have in common are that they are small and that they are game. This is exemplified by the two predator-prey simulations presented in this paper.

Simulation I

The first simulation was designed to address two central questions. First, what are the relative rates of population growth among these taxa, and second, at what level can they sustain continued predation by humans? Only a brief description of this model is provided here, but a more detailed portrayal is available elsewhere (Stiner et al. n. d.). Each species is simulated as an actual set of individuals. Each individual is characterized by an age and sex, but females are also assigned a next age of reproduction based on the age of reproductive maturity and birth spacing. Population dynamics are controlled by a series of fertility and mortality parameters (Table 4.1); values for these parameters were obtained from wildlife and population studies (see Stiner et al. n.d. for references). It was often necessary to broaden the taxonomic scope within genera to adequately obtain parameter values. To account for variability in population growth rates seen in natural populations, it was necessary to create a high growth model (HGM) and a low growth model (LGM) for each taxon. These are intended to represent extremes of population growth with typical dynamics certainly falling somewhere in between. Hunting is controlled by two parameters: annual kill percentage, and minimum age (or size in the case of tortoises) to hunt. Annual kill percentage is simply a fixed percent of the population to be removed by hunting every year, and minimum age or size is a threshold above which individuals are taken preferentially. This allows the simulation of age- or size-specific culls. Age- and/or size-based selectivity might be expected, especially for small game, in that returns are low for small package size resources.

The simulations begin with an initial population of 15 individuals. This is allowed to grow until carrying capacity, which is arbitrarily set at approximately 900 for the HGM and 400 for the LGM. Population equilibrium is achieved by allowing juvenile mortality rates to be density dependent. Once the populations reach carrying capacity, they are hunted at a fixed annual percentage focusing on adults until no adults remain; then, juveniles were hunted. If the population goes extinct before 200 years, the annual culling rate is considered to be unsustainable, otherwise it is considered sustainable.

Table 4.1. Fertility and mortality parameter values for tortoises, hares and partridges for high (HGM) and low growth models (LGM) of Simulation I.

Parameter	Tortoises		Partridges		Hares	
	HGM	LGM	HGM	LGM	HGM	LGM
Fertility						
Age of first female reproduction (yrs)	8	12	0.75	1.0	0,75	1.0
Birth spacing (days)	365	730	365	365	365	365
Minimum no. of offspring*	7	7	9	7	11	9
Maximum no. of offspring*	14	14	11	9	13	11
Mortality						
Maximum life span (yrs)	60	60	12	12	8	8
Age of onset of adult mortality (yrs)	1	1	0.5	0.5	0.2	0.2
Annual adult mortality	0.053	0.093	0.4	0.5	0.5	0.6
Annual juvenile mortality	0.7	0.85	0.6	0.7	0.42	0.6

*per reproductive episode
See Stiner et al., n.d. for citations to wildlife literature used to estimate values

Results Of Simulation I

Hares and partridges consistently behave very similarly with high growth rates and high hunting tolerances, while the tortoises show precisely the opposite trends. In the HGM, the hares grow from a population of 15 individuals to almost 900 in less than ten years, and partridges mimicked this almost exactly (Figure 4.1). In the tortoise HGM, the population takes approximately 50 years to do the same thing. The difference is even greater in the LGM (Figure 4.1). Slow versus fast population growth has clear consequences for a population's sensitivity to hunting pressure; rapid population growth translates to high resiliency, while slow growth means high susceptibility.

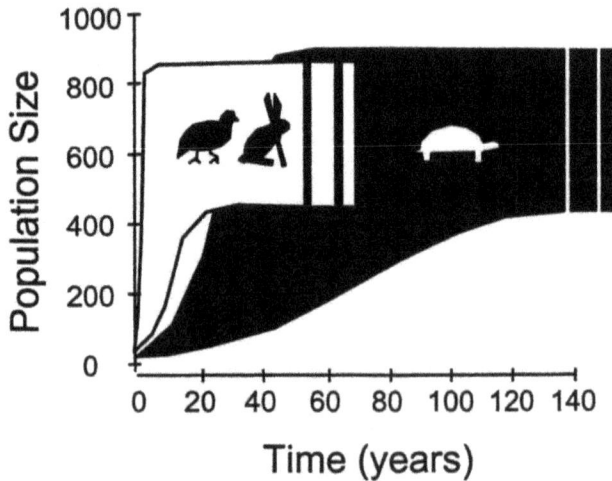

Figure 4.1. Population growth curves for tortoises, hares and partridges from Simulation I for the HGM and LGM. The lower limit of each curve denotes population growth in the LGM and the upper limit denotes population growth in the HGM. Because the hare and partridge curves were virtually identical, they are depicted as a single curve. In actuality, partridge growth rates slightly exceeded those of hares.

In the tortoise LGM, the population can only sustain up to a 3% annual cull per year. Under high population growth conditions in the HGM, they can withstand up to an 8% cull (Figure 2). As is expected, hares are very resilient to human exploitation. In the LGM, the population remains viable after an 18% annual cull (Figure 2). In the HGM, the population goes extinct if the culling rate exceeds 50% per year. Chukar partridges in the LGM can sustain as much as a 20% population loss per year, while in the HGM, amazingly they can withstand up to a 65% annual cull without being driven to extinction (Figure 4.2).

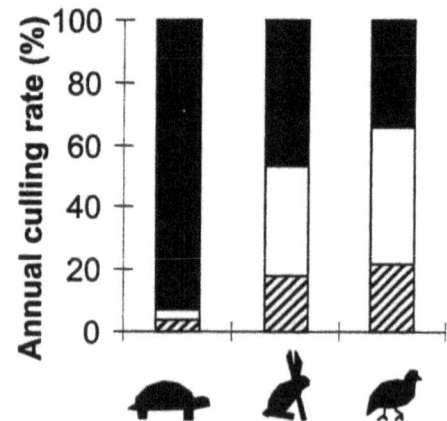

Figure 4.2. Hunting thresholds for tortoises, hares and partridges from Simulation I for the HGM and LGM. The range of sustainable annual culling rates for the LGM is shown hatched and for the HGM in white. Unsustainable culling rates are shown in black. A culling rate was considered sustainable if the population remained viable after 200 consecutive years of exploitation.

These results are by no means crisp in that they delimit rather large ranges of sustainable annual culling and population growth rates, but the value of such "fuzzy" results is that they place conservative limits on predator-prey relations. Surprisingly, even with a large degree of slop in the data, there is clear separation between tortoises and their warm blooded counterparts. This disparity should have clear archaeological consequences. That is, with increasing intensity of occupation, all things being equal, tortoise populations should decline more rapidly than those of hares and partridges due to human exploitation. Through time, this should result in an increase in the number of hares and partridges in the archaeological record as tortoises decline. A second model was created to test this proposition.

Simulation II

In this simulation, a group of foragers returns to a site every year and exploits the surrounding "catchment area". Instead of simulating each species separately, they were combined into a single "ecosystem" model. This abstract ecosystem contains only tortoises, partridges, and hares. The advantage of modeling the populations this way is that the frequency of one species in the ecosystem will affect to what extent the others are exploited. Theoretically, this should only be true of highly ranked taxa because no matter how frequent a low ranked species is, it may never be used; but if we set this basic tenet of optimal foraging aside for a moment, it becomes easier to understand the dynamics of the system. It is important not to take the results of the following model too literally. It is intended only to be a heuristic device for exploring how each taxa responds to human exploitation in an ecosystem. The general trends are intended to be the take home point; the specific results are not.

To model a large number of individuals of three taxa simultaneously, it was necessary to change the format of the model to make it run more efficiently. Therefore, populations are modeled as matrices; each cell in a matrix contains the number of individuals in a particular age and sex cohort. The dynamics of the model are controlled by life table matrices which store age- and sex-specific annualized fertility and mortality data. Using this format, it is possible to model very large populations and to do so very quickly. Instead of using two sets of fertility and mortality data for each taxon as was done with the HGM and LGM in Simulation I, a mean model was used which fell half way between the HGM and LGM from Simulation I (Table 4.1).

In this ecosystem, it was necessary to estimate relative densities of tortoises, hares, and partridges. Unfortunately, there is a lot of variability reported in these values in the wildlife literature, and very little data from the study area in question. Also, it is unclear how comparable interspecific density estimates are, due to different censusing techniques. It was assumed, however, that tortoises exist at higher densities than hares and partridges due to lower metabolic rates and predation levels. Tortoise populations were allowed to grow to equilibrium at 5000 individuals. Partridges and hares were allowed to grow to half of that at 2500 individuals. These values are somewhat arbitrary, but really have no affect on the outcome of the simulation because they are held constant for all runs.

Three variables are used to model occupation intensity: human group size, duration of site occupation, and return rate. Group size is the number of individuals occupying the site. Duration of occupation is the number of days the site is occupied during one year, and return rate is the number of animals taken per person per day. The product of these determines the gross number of animals to be acquired per year. For all of the runs presented here, group size is held constant at 25, and return rates are held constant at 0.1 small game animals per person per day. Therefore, each day, 2.5 small game animals are removed from the natural populations by hunting. The only variable that is adjusted is occupation duration which is incremented from 1 to 365 days in 8 steps of 52 days. Small game species are taken according to their actual frequencies in the environment. The simulation records the natural population levels of tortoises, hares, and partridges in the environment, and the frequencies at which they enter the archaeological record for a period of 100 years.

Results Of Simulation II

Simulation II indeed confirms that tortoises should become less frequent with increasing duration of occupation (Figure 4.3). Again, the hares and partridges behaves very similarly. For a 53-day occupation length, the system remains in equilibrium. Tortoise, partridge, and hare populations in the ecosystem remain relatively stable. Tortoises comprise about 50% of the archaeological record, while partridges and hares comprise approximately 25% each. At a 105-day occupation length, tortoise frequencies begin to fall in the ecosystem and in the archaeological record. Partridge and hare populations, however, stay relatively undisturbed in the ecosystem, while they become slightly more prevalent in the archaeological record. With a 157-day occupation, the tortoise population drops rapidly, while hare and partridge populations slowly decline. In the archaeological record, partridges and hares become more common than tortoises after 50 years as the hares and partridges also become more common in the ecosystem. When the occupation duration is increased to 219 days, all 3 species are driven to extinction as their sustainable cull rates are exceeded in year 57. Notice again that the tortoise population drops rapidly, while hares and partridges decline much more slowly until tortoises are so rare that hares and partridges are exploited almost exclusively. Because at this point hare and partridge populations are already depressed, they are driven to extinction by the increased burden they are forced to carry. Also notice that hares and partridges become numerically dominant archaeologically in 30 years, 20 years faster than in the previous run. Finally, when duration of occupation is increased to 365 days, or continuous occupation, there is an exaggeration of the patterns seen in the previous run with hares and partridges dominating the archaeological record after only 18 years, and total extinction occurring in year 22.

Discussion

In actuality, extinction never occurs for any of these taxa. There are many ways in which the simulation may be unrealistic in this respect, but none of these will alter the final outcome. Predation may not have been as intense or as consistent as it has been modeled. Prey population sizes within the site catchment area may have been larger. Also, small game populations likely had years between predation episodes in which their populations could recover. As hares and partridges would bounce back more quickly than tortoises, the net result of increased occupation intensity, even with predation hiatuses, should be the same. Also, it is likely that as population densities of small game drop to low levels, small game largely drop out of the diet as humans are unable to find them. Lastly, local extinctions may have occurred in the close vicinity of base camps while

Entering the Archaeological Record

Living in the Ecosystem

53 day occupation

105 day occupation

157 day occupation

209 day occupation

261 day occupation

313 day occupation

365 day occupation

Frequency

Population Size

Time (years)

Time (years)

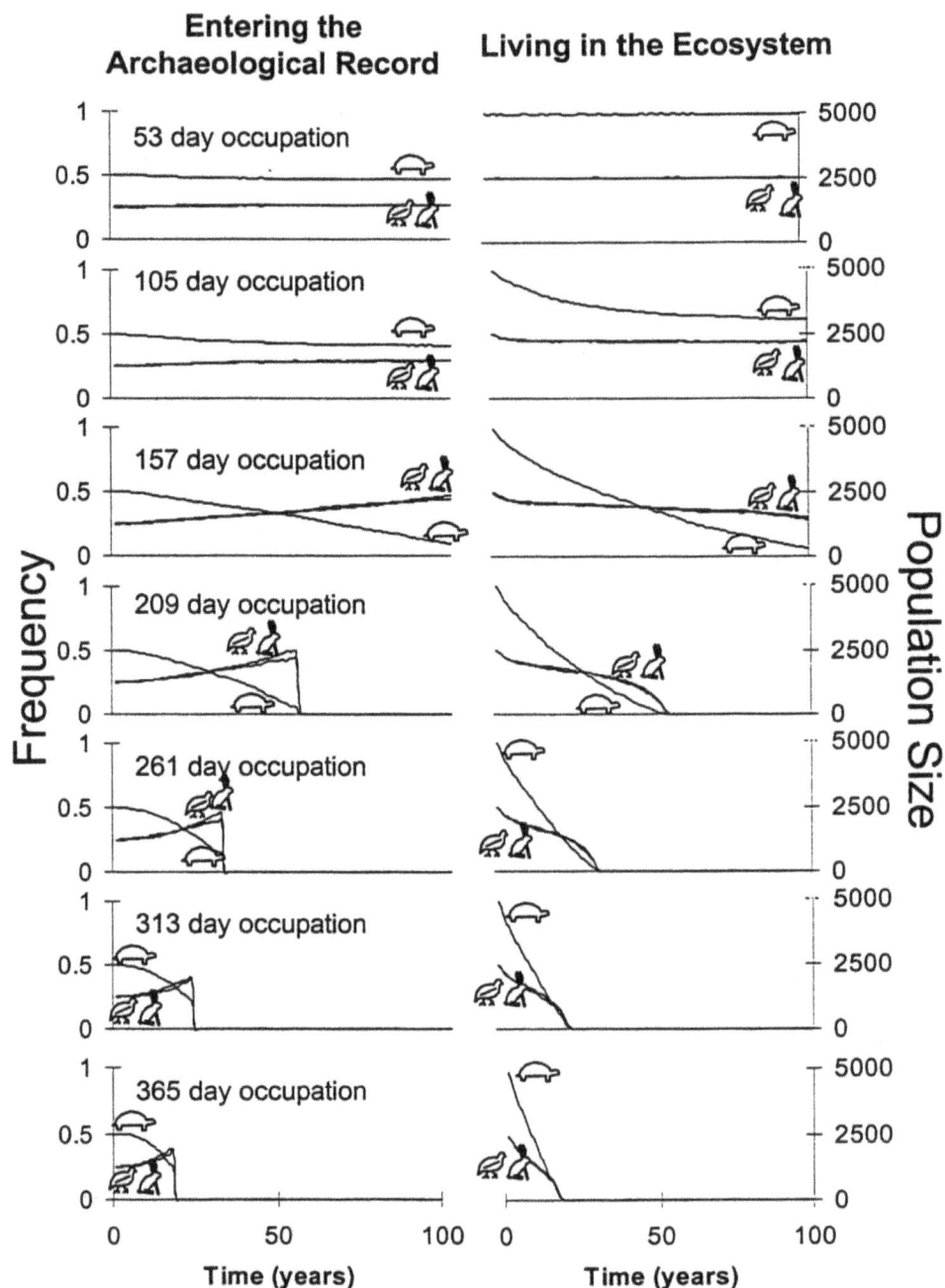

Figure 4.3. Results of Simulation II showing the frequency of tortoises, hares and partridges entering the archaeological record (left) and the numbers of animals in the ecosystem (right) for 100 years. Each pair of graphs represents a particular length of site occupation duration ranging from 53 days per year (top) to 365 days per year (bottom).

other populations persisted in buffer areas between base camps and/or band territories.

Thus, it is clear that with increasing duration of occupation, greater numbers of hares and partridges, relative to tortoises, are expected to enter the archaeological record. This results from differences in reproductive potential of these small game species. But is it justifiable to forsake the principals of optimal foraging? In Simulation II animals are exploited in proportion to their natural frequencies in the environment. In fact, it is known that hunter-gatherers do not forage in this fashion. Instead, they tend to maximize foraging returns by utilizing only those resources that are

most cost efficient (e.g. Belovsky 1987; Hawkes et al 1982; O'Connell and Hawkes 1981; Smith 1983). How would tortoises, hares, and partridges be ranked in an optimal foraging framework, and how would this affect the outcome of the model?

Fortunately, the answers to these questions are relatively straightforward. In general, small game tend to be low ranking due to low return rates, but tortoises do not obey this generalization. While tortoises do come in small packages, they require virtually no energy to capture. Thus, it is expected that foragers would find it difficult to pass up the opportunity to collect tortoises. The only real cost in using tortoises as a food resource is finding them, which can occur during the normal course of foraging. Partridges and hares, on the other hand, have developed sophisticated defense mechanisms because they must elude multiple natural predators. Hares tend to rely on speed and subterranean burrows, while partridges have flushing and flight. Therefore, for humans, the acquisition of small, fast, agile prey is considerably more expensive because it often involves social cooperation and/or elaborate technologies. In this sense, tortoises are expected to have been very highly ranked, while hares and partridges should have been ranked lower. The implication is that chronologically, tortoises should have been added to the diet very early on, and should decline very quickly with increasing intensity of occupation as they seem to do (Munro this volume; Stiner et al. 1999 n. d.). In contrast, partridges and hares, should be added later and should increase in frequency with greater occupation intensity as higher ranked species, such as tortoises, become less available.

This trend away from tortoises to hares and partridges, however, may not simply be a function of reduced residential mobility (*sensu* Binford 1980) and increased length of occupation of sites. As modeled, this could just as easily result from larger group sizes and/or higher small game return rates. Distinguishing between these possibilities requires the consideration of multiple independent lines of evidence. Other possible interfering factors are technological innovations that may affect the relative rankings of any of these species, and environmental change that could selectively favor certain taxa. Therefore, although the model portrays a fairly simple system, actually implementing the relative frequencies of tortoises, hares, and partridges as a measure of occupation intensity is not so straightforward. If, however, environmental and technological change can be ruled out or accounted for, the potential payoffs are large because such data can be collected across regions to assess the relative intensity of exploitation of ecosystems in space and time. This is effectively a method for exploring population pressure or intensification mosaics across landscapes. As population disequilibria within regions have been proposed as prime movers in the agricultural transition (e.g. Binford 1968), this method could provide a partial test of such ideas.

In order to perform such an analysis, two further issues must be addressed, namely the state of the ecosystem "inherited" by the foraging group moving into an area, and time averaging. From Figure 4.3, it is clear that when a

"pristine" ecosystem is encountered, tortoises will always dominate the assemblage at first, no matter how intensive an occupation may be. It is only as the site is occupied or reoccupied over multiple years that hares and partridges will increase in frequency in the archaeological record. Similarly, long term occupations may not necessarily be dominated by fast reproducing prey if they are punctuated by long occupation hiatuses in which the ecosystem is allowed to bounce back to its virgin state where all prey species have reattained equilibrium at carrying capacity. This, however, should not be viewed as a disadvantage of this method. In fact, the archaeological record is generally suitable to such analyses in that ephemeral "time slice" occupations are the exception, not the rule. Also, as human population densities increase, the probability of inheriting an ecosystem in a somewhat depressed state is also increased which further enhances the sensitivity of the model to monitoring human demography.

Finally, small animals have been largely pushed to the wayside by archaeologists. They tend to find their way on to species lists and into paleoclimatic reconstructions, but discussions of their relevance to archaeological research usually ends there. This paper has shown one way in which small game species can inform us about issues that go far beyond dietary breadth and paleoecology. Clearly the variation in life history discussed for these taxa can provide us with a rough measure of occupation intensity at archaeological sites. This results from the simple fact that as constituents of ecosystems, human beings shape those ecosystems relative to the intensity at which they exploit them. In turn, prehistoric humans provide us with a sample of that ecosystem in the form of an archaeological record. The key to extracting inferences from the archaeological record is in understanding the dynamics of that interaction.

Acknowledgements
This paper has resulted from collaboration with Mary Stiner and Natalie Munro. Their help, guidance, and support was fundamental in the development and testing of the models presented in this paper. I would also like to thank Steve Kuhn for aid during mathematical and statistical crises, and Jon Driver for the opportunity to participate in this symposium. Finally, I thank Kaya T. Dog for love and companionship through it all.
The research reported in this paper was funded by a National Science Foundation grant to Mary C. Stiner [SBR-9511894].

References

Bar-Yosef, O. and R. H. Meadow, 1995. The origins of agriculture in the Near East. In *Last Hunters First Farmers*, ed. T. D. Price and A. B. Gebauer. Santa Fe: School of American Research Press, pp. 39-94.

Belovsky, G. E., 1987. Hunter-gatherer foraging: A linear programming approach. *Journal of Anthropological Archaeology* 6:29-76.

Bender, B., 1978. Gatherer-hunter to farmer: a social perspective. *World Archaeology* 10:204-222.

Binford, L. R., 1968. Post-Pleistocene adaptations. *In New Perspectives in Archaeology*, ed. S. R. Binford and L. R. Binford. Chicago: Aldine, pp.313-341.

Binford, L. R., 1980. Willow smoke and dogs' tails: hunter-gatherer settlement systems and archaeological site formation. *American Antiquity* 45,4-20.

Bonner, J. T., 1965. *Size and Cycle: An Essay on the Structure of Biology.* Princeton: Princeton University Press.

Charnov, E. L., 1993. *Life History Invariants: Some Explorations of Symmetry in Evolutionary Ecology.* Oxford: Oxford University Press.

Childe, V. G., 1928. *The Most Ancient East*. London: Routledge and Kegan Paul.

Cohen, M. N., 1977. *The Food Crisis in Prehistory*. New Haven, Connecticut: Yale University Press.

Flannery, K. V., 1969. Origins and ecological effects of early domestication in Iran and the Near East. In *Man, Settlement, and Urbanism*, ed. P. J. Ucko and G. W. Dimbleby. Chicago: Aldine, pp. 73-100.

Hawkes, K., K. Hill and J. F. O'Connell, 1982. Why hunters gather: optimal foraging and the Ache of eastern Paraguay. *American Ethnologist* 9:379-398.

Hayden, B., 1981. Research and development in the Stone Age: technological transitions among hunter-gatherers. *Current Anthropology* 22, 519-548.

Hayden, B., 1990. Nimrods, piscators, pluckers, and planters: the emergence of food production. *Journal of Anthropological Archaeology* 9, 31-69.

O'Connell, J. F. and K. Hawkes, 1981. Alyawara plant use and optimal foraging theory. In *Hunter-Gatherer Foraging Strategies*, ed. B. Winterhalder and E. A. Smith. Chicago: The University of Chicago Press, pp. 99-125.

Pianka, E. R., 1970. On r- and K-selection. *The American Naturalist* 104, 592-597.

Smith, E. A., 1983. Anthropological applications of optimal foraging theory: a critical review. *Current Anthropology* 24, 625-651.

Stiner, M. C., N. D. Munro and T. A. Surovell, In prep. The tortoise and the hare: small game use, the broad spectrum revolution, and Paleolithic demography.

Stiner, M. C., N. D. Munro, T. A. Surovell, E. Tchernov and O. Bar-Yosef. 1999. Paleolithic population pulses evidenced by small animal exploitation. *Science* 283, 190-193.

Wright, H., Jr. 1977. Environmental change and the origin of agriculture in the Old and New Worlds. In *Origins of Agriculture*, ed. C. A. Reed. The Hague: Mouton, pp. 281-318.

SMALL GAME AS INDICATORS OF SEDENTIZATION DURING THE NATUFIAN PERIOD AT HAYONIM CAVE IN ISRAEL

Natalie Munro

Department of Anthropology, University of Arizona, Tucson, AZ, 85721, U.S.A.

Introduction

Demographic pressure has repeatedly been cited as a powerful instigator of economic and social change at many critical junctures in prehistory. Rising population densities have been proposed as the driving mechanism behind the process of sedentization and the adoption of agriculture in many parts of the globe. The best known example of sedentization occurred between 12,000 and 8,000 years ago when mobile hunter-gatherers in Southwest Asia settled into permanent agricultural villages. Although agricultural origins have been extensively researched, the causal mechanisms that led to the adoption of agriculture remain unclear. Better understood, however, are the necessary prerequisites (reviewed by Gebauer and Price 1992; Keeley 1995; Price and Gebauer 1995): social circumscription, a tendency toward social complexity, a rich and diverse subsistence base, and most importantly demographic pressure and sedentism (e.g., Boserup 1965; Binford 1968; Cohen 1977; Flannery 1969; Rosenberg 1998). To shed further light on the process of sedentization and the Neolithic transition in Southwest Asia this paper examines human economic adaptations at Hayonim Cave in Israel during the Natufian period between about 13,000 and 11,000 years ago, just prior to the advent of agriculture.

For the past 30 years Flannery's (1969) Broad Spectrum Revolution (BSR) hypothesis has been the favoured explanation for dietary change in the Late Pleistocene. This hypothesis proposes that the addition of a diverse array of low-utility plant and animal resources was a symptom of population pressure in Southwest Asia (see also Binford 1968). Traditionally, the BSR has been interpreted as a response to the reduced availability of, or access to, large high-quality resources that potentially resulted from high human population densities (Bar-Yosef and Belfer-Cohen 1989, 1991). The sustained use of environments by humans exerts pressure on resources and may eventually lead to the depression of local plant and animal populations (Broughton 1994; Speth and Scott 1989). Humans must adapt their strategies to maximize foraging efficiency in response to self-induced disturbances in resource distribution and abundance. Recently, a number of researchers using diversity indices to measure species richness (number of species) or evenness (number of species and their relative abundance) in an assemblage over time have begun to challenge this long-standing hypothesis. These diversity measures have been applied to a range of Levantine faunal assemblages dating to the Middle Paleolithic through the Neolithic periods and beyond (Edwards 1989; Horwitz 1996; Kislev et al. 1992; Neeley and Clark 1993). The results of these tests suggest that if hominids have been using broad spectrum economies, these strategies have been in place since at least the Middle Paleolithic period. A diversity study conducted by Neeley and Clark (1993) exemplifies this apparent contradiction: although it failed to detect increased diet breadth in the Late Pleistocene, it did demonstrate that Natufian assemblages are much different from earlier assemblages. Unfortunately, the

nature of these changes remain unclear. Thus, it appears that while the BSR hypothesis touches on a real phenomenon, evaluation of the hypothesis requires a new approach.

Many researchers agree that generalist subsistence economies were in place by the transition to agriculture, but debate about just how broad the resource base of the terminal Paleolithic really was continues (Bar-Yosef and Belfer-Cohen 1989, 1991; Edwards 1989; Henry 1989; Neeley and Clark 1993). Moreover the need to restructure the BSR hypothesis must not obscure the fact that economic change did occur in Southwest Asia at the end of the Pleistocene. Accordingly, we must now ask what is the nature of this change, and is it a response to increased demographic pressure as Flannery suggested?

Although it figures almost universally in models explaining agricultural origins, the existence of population pressure has not been conclusively documented in the archaeological record. Researchers have depended primarily on indicators of sedentism as indirect evidence for increasing human population pressure during the Natufian period (e.g., Bar-Yosef and Belfer-Cohen 1989; Belfer-Cohen 1991; Byrd 1989; Henry 1989; Tchernov 1984, 1991, 1993). The remains of stone structures, dense deposits of cultural remains, heavy grinding stones, and commensal animals have commonly been offered as evidence that at least some Late Epipaleolithic human groups were semi-sedentary hunter-gatherers. Though this evidence of increased sedentism is instructive, it remains largely circumstantial as a measure of human demographic change. Other more conclusive archaeological correlates must be developed to monitor the intensity of human population pressure during the Late Pleistocene.

Small Game and Human Demography

Research on economic change in the Late Pleistocene in Southwest Asia has focused largely on plants and large animal species (Byrd 1989; Colledge 1991; Cope 1991a, 1991b; Campana and Crabtree 1990; Davis 1978, 1983; Hillman 1996; Hillman et al. 1989; Hopf and Bar-Yosef 1987; Lieberman 1991, 1993; Unger-Hamilton 1991). Gazelle have been of primary interest because they are the most common species in Epipaleolithic assemblages, and they contributed the greatest quantity of meat to human diets. Recently, attention has been drawn to the explanatory power of small game animals for addressing questions of demographic change (Stiner et al. 1999, n.d.). In most Paleolithic assemblages the large game component is entirely composed of ungulate species. Because these species exhibit relatively uniform responses to predator pressure, they provide little information on human demography. Alternatively, there is great diversity in the responses of small game taxa (e.g., tortoises, hares, birds) to predation pressure. These taxa differ in such potentially informative characteristics as escape strategy, population density, and reproductive strategy. Thus, although small game have been a largely unstudied component of Paleolithic assemblages (but see Davis 1989; Davis et al. 1994; Pichon 1984, 1991; Tchernov 1984, 1991,1993), they can greatly inform our interpretations of human demography and site occupation intensity.

The BSR can also be tested by considering the reproductive and behavioral characteristics of species represented in a given assemblage, and, more specifically, by evaluating the demographic resilience of the most common species to intensive human hunting pressure. It may be the types of animals that dominate the Natufian diet, rather than the number of species, that are most likely to inform us about economic and demographic change over the course of the agricultural transition (see also Stiner et al. 1999, n.d.). Diversity indices provide useful measures of the number of different taxa in archaeological assemblages; yet, because they cannot isolate the biological correlates of species composition, additional steps need to be taken. Here, the reproductive and behavioral characteristics of small animal species are presented as independent referents of human demography and settlement occupation duration during the Late Pleistocene in the Levant. This aspect of the research builds on predator-prey simulations presented by Surovell (this volume).

Optimization models assume that a forager's goal is the maximization of reproductive success. This notion is generally operationalized using energy as a currency (Kelly 1995; Krebs et al. 1983; Perry and Pianka 1997; Schoener 1986; Stephens and Krebs 1986). Foraging theory predicts that hunters will choose only the highest-ranked resources if those resources provide higher average energy returns than the diet also including lower-ranked resources (Emlen 1966; MacArthur and Pianka 1966; Stephens and Krebs 1986). The abundance of high-ranked prey species determines whether a forager will need to add lower-ranked resources to its diet. For human hunters, the highest-ranked resources are usually large game animals (e.g., ungulates), which provide high caloric returns per individual due to their large body size. If large game are hunted intensively, their populations may eventually become depressed, and necessitate the addition of lower ranked resources to the diet. Although small animals provide much less energy per package, they can still be highly-ranked if they require little handling time to procure (e.g., tortoises). Small animals with rapid escape strategies (e.g., hares and birds) are much more energetically expensive, because they provide low energy returns and have high handling costs. These resources are not expected to be included in human diets unless higher-ranked resources are sparse, perhaps as a result of intensive human hunting pressure.

Clearly, it is not economical to capture small fast game using techniques developed for the procurement of large game. Small fast animals are efficiently captured using technologies that target multiple individuals (e.g., nets) or those that do not require active pursuit of the prey (e.g., traps and snares). It is generally agreed that new technologies including nets, traps, snares and the bow and arrow were in use by the Late Pleistocene (e.g., Bar-Yosef and Belfer-Cohen 1989; Hayden 1981; Soffer 1985), though their residues rarely preserve in the archaeological record.

These technologies make the capture of otherwise expensive small game feasible, and increase the potential energetic yield that human hunters can extract from a unit of territory. Technology also has the potential to change the relative ranking of resources. For example, if the use of a technology reduces the handling costs required to capture a resource, the net gains from that resource will increase, and may exceed the net gains from a previously higher-ranked resource. Nets and traps may increase the potential net energetic gains from a unit area, but the costs required to manufacture, maintain and operate these technologies are high (Bailey and Aunger 1989; Bousman 1993). Though the innovation of these new technologies is expected to make the capture of small game worthwhile, it is argued that due to the high handling costs of these technologies, the capture of large quantities of small game may still result in an overall reduction in foraging efficiency. The use of nets, traps and snares is therefore not expected to significantly alter the ranking of species such as hares and birds in comparison to ungulates and tortoises.

The number of days a human group occupies a residential site directly impacts the distribution and abundance of prey taxa in the surrounding environment. When human populations occupy a site for long periods, they require more resources, and place increased demands on prey populations. When foraging, a predator initially selects only the highest-ranked prey type if the preferred prey is sufficiently abundant to meet its daily energy requirements. The prey population can maintain its population size until the pressure of sustained harvest by the predator exceeds the intrinsic rate of increase of the prey population. At this point, the prey population begins to decline. Sustained hunting of a shrinking prey population will accelerate its depletion. By depressing a prey population, particularly a preferred one, human hunters are forced to revise their hunting strategies so that they may continue to maximize foraging returns. The hunter must take a reduction in foraging efficiency and add lower-ranked species to its diet. Thus, the sustained hunting of a habitat for long periods is expected to correlate with a reduction in human foraging efficiency. Foraging efficiency can be measured as the proportion of high-ranked to low-ranked species in the diet of a human group, and is manifested archaeologically as the proportions of animal taxa in faunal assemblages.

The availability of prey in an environment also depends on the resilience of a prey taxa to predation. An animal's reproductive strategy determines how many individuals can be harvested before its population begins to decline. Stiner et al. (1999, n.d.; see also Surovell this volume) predict the effects of intensified human hunting on three small animal species: tortoise, hare, and partridge. These simulations clearly illustrate taxonomic differences in prey resilience and recovery rates in response to varying percentages of hunting intensity. The results show clear differences between the population dynamics of the slow reproducing tortoise and those of the much faster reproducing hare and partridge. Hares and partridges have high population turnover rates and short life spans, and they reach reproductive maturity at a young age. Their populations are thus able to withstand high levels of hunting intensity. Tortoises and ungulates fall on the other side of the reproductive spectrum. They have low population turnover, reach reproductive maturity slowly, and because they have longer life spans, are able to reproduce for many years. These populations are far more easily captured and much more heavily affected by human predation. Following the arguments outlined above, Stiner et al. (1999, n.d.) are able to link changes in the proportions of small game species in Paleolithic assemblages from Hayonim Cave in Israel to increasing human population densities. The observed shift from slow-reproducing to fast-reproducing small animals in the assemblage over time is provocative but requires a fuller examination in Natufian assemblages.

Intensification of human hunting in an ecosystem may result from either an increase in regional human population density, or an increase in the length of occupation at a habitation site, or both. Because only one site is the subject of this study, human demography cannot be measured on a regional level; however, it is possible to evaluate the intensity of site occupation at Hayonim Cave. When humans move their residential camps frequently, they are able to harvest required resources without significantly impacting the population size or growth potential of local prey populations. The faunal assemblages at small sites are therefore expected to be dominated by high-ranked species. An increase in the length of occupation of the residential camp requires an increase in the amount of resources that must be harvested from the local ecosystem. As site occupation duration increases, populations of high-ranked prey taxa are expected to become depressed, and will be captured with progressively less frequency as site occupation lengthens. The assemblages from small (shorter occupations) and large sites (longer occupations) inhabited during the Natufian period are therefore expected to display differential patterns of taxonomic representation. Sites occupied for short periods are expected to exhibit greater proportions of high-ranked prey taxa than those occupied for longer periods.

The intensity of human hunting during the Natufian period is evaluated here using the arguments outlined above and the predictions generated by the model presented by Surovell (this volume). These simulations model the impact of human hunting on populations of common small game species for site occupations of varying duration. The same three prey species are hunted as in previous simulations (cf. Stiner et al. n.d.), but the prey are contained within a single ecosystem, rather than existing as independent units. The simulated hunters pursue prey types upon encounter without discriminating between them. During each iteration, the prey populations experience natural population growth and are subject to human hunting. The simulation tracks the proportions of taxa in the ecosystem as well as those that are deposited in the archaeological record (hunted individuals) as refuse during each cycle of the model (see Surovell this volume for a detailed explanation).

The simulations predict that as site occupation duration increases, the number of tortoises in the ecosystem will decline in response to predator pressure. Tortoise populations are unable to recover from sustained hunting episodes and are quickly driven to extinction. Hare and partridge populations, on the other hand, behave much

differently. These populations are able to endure much longer periods of hunting each year before they become depressed, since their populations are able to bounce back rapidly during the part of the year when they are not hunted, as a result of their high population growth rates. Thus, as the length of site occupation increases, the number of tortoises in the environment relative to hares and partridges decreases.

Evaluation of archaeological data from the Pleistocene/Holocene transition in light of these predictions allows for an estimation of the relative intensity of site occupation. Although, Surovell's predictions are not intended to be applied quantitatively, they do enable the identification of trends or changes in the intensity of site occupation over time or space. The model predictions are considered here by examining the proportions of different game species recovered from Hayonim Cave, with the assumption that an increase in hunting intensity constitutes an increase in the length of occupation of human sites.

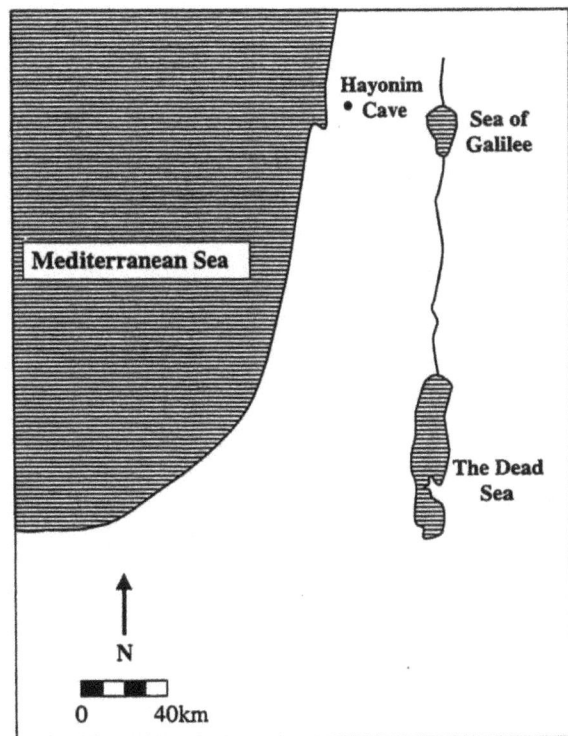

Figure 5.1. Map of the Levant, showing the location of Hayonim Cave

Hayonim Cave Natufian Fauna

Hayonim Cave is a multicomponent site located in the Mediterranean biogeographic zone in northern Israel (see Figure 5.1). Multiple occupational layers at the site span at least 200,000 years, with intermittent hiatuses. Cultural layers from the Mousterian, Aurignacian, Kebaran and Natufian layers provide an ideal series for observing economic change throughout the late Pleistocene until the Holocene boundary, which is synchronous with the Neolithic Revolution in the Near East. Fortunately, the Natufian component is exceptionally large (Bar-Yosef 1991; Belfer-Cohen 1988, 1991). To date, much of the fauna from the earliest layers in the cave has been analyzed, and the data are available for comparative studies (Rabinovich 1998; Stiner and Tchernov 1998, Stiner et al. n.d.).

The faunal assemblages from all periods at Hayonim are characterized by excellent preservation and abundant occupational records. The Hayonim project personnel invested in thorough recovery which was accomplished by means of point proveniencing in three dimensions, wet and dry-screening, and "picking" through screened sediment to recover fauna that were initially overlooked. These strategies have enabled the recovery of a full size spectrum of animal taxa, including the tiny remains of small animals that tend to pass through the mesh or go unnoticed during screening.

Though the primary excavations in the Natufian layer at Hayonim Cave took place during the 1970's (Bar-Yosef 1991; Bar-Yosef and Goren 1973), a research team directed by Ofer Bar-Yosef returned to Hayonim Cave in 1992 to excavate the Mousterian deposits. In conjunction with this study, small-scale excavations were reopened in the Natufian layer. The goal of this exposure was the documentation of two structures rich in architectural, faunal, and technological material. These ongoing excavations have already yielded a sizable sample of fauna which is the central subject of this discussion. In this analysis, only the microfauna (small rodents, Passerine birds and amphibians) are excluded from the assemblage. At Hayonim Cave, the bones of microfauna are often recovered as complete or nearly complete skeletons, and have a fresh looking appearance in comparison to elements of larger species. The cave ceiling above the Natufian deposits contains several niches which are home to modern day barn owls (Tyto alba), which regurgitate pellets rich in rodent and small bird bones (Andrews 1990). It is likely that owls also occupied the cave in prehistory, as the bones of Tyto alba, and several other owl species have also been recovered from the Natufian layer (Pichon 1984). Finally, microfauna in the Natufian layer rarely exhibit taphonomic evidence typically associated with economic use, such as burning and breakage. Thus, the microfauna are excluded from the analysis as they appear to have been introduced to the cave by owls or represent natural deaths.

Results

The Natufian faunal assemblage from the new excavations at Hayonim Cave contains 2,876 identifiable specimens representing at least 27 species (see Table 5.1 for summary). All of the percentages and absolute numbers provided here are NISP counts as this measure is believed to be the most appropriate quantitative method for examining general trends in species relative abundance (Grayson 1984; Lyman 1994).

Table 5.1. NISP counts from the Natufian component of Hayonim Cave

TAXON	NISP
Reptilia	
Ophisaurus apodus	9
Agama stellio	65
Testudo graeca	513
Total Reptilia	587
Pisces indeterminate	3

Aves	
Aquila chrysalis	3
Buteo buteo	22
Falconiformes	21
Vulture	2
Galliformes	1
Alectoris chukar	231
Coturnix coturnix	1
Strigidae	4
Anser anser	1
Fulica atra	8
Columbiformes	1
Columba livia	2
Small Aves	14
Medium Aves	156
Large Aves	36
Huge Aves	6
Indeterminate	1
Total Aves	511
Small Mammals	
Erinaceus europeaeus	4
Sciurus anomalus	29
Lepus capensis	444
Total Small Mammals	477
Carnivora	
Vulpes vulpes	23
Canis sp.	2
Martes foina	2
Meles meles	1
Mustelidae	4
Panthera pardus	5
Felis sp.	26
Indeterminate	22
Total Carnivora	85
Ungulates	
Sus scrofa	14
Capreolus capreolus	4
Dama mesopotamica	19
Cervus elaphus	8
Cervidae	17
Gazella gazella	565
Capra ibex	6
Bos primigenius	7
Small ungulate	501
Medium ungulate	59
Large ungulate	16
Total Ungulates	1216
Total NISP	2876

Analysis of the Natufian fauna from Hayonim Cave reveals two noteworthy changes from earlier occupations at the site. First, there are unprecedentedly high proportions of small animal taxa that have been demonstrably modified by humans: small game elements in Natufian sites are highly fragmented (mean fragment size is 2.0 cm), and many tortoise (*Testudo graeca*) carapace and plastron fragments exhibit cone fractures, most likely gained when struck by a hard object. Burning was also common across both large and game taxa in the Natufian layer (25% of total NISP), but was especially prevalent on small mammal bones, most notably hares (48%). Burned small mammal bones are distributed over a range of contexts and are interspersed with non-burned bones suggesting that burning took place prior to deposition. The ungulates continue to comprise a significant proportion of the Natufian fauna (42%), particularly in light of their large body size, and are composed nearly entirely of gazelle (*Gazella gazella*). However, the small game are also abundant (55%), even surpassing the ungulate fauna in sheer numbers. Although they have much smaller body sizes, the investment required to capture such large numbers of small animals is substantial. As in most of the early layers carnivores represent only a small fraction of the identifiable specimens (3%). Compared to earlier periods, small game comprise a much higher proportion of the Natufian assemblage (see Figure 5.2). Earlier assemblages are composed of much greater proportions of ungulates, including several larger bodied species such as fallow deer (*Dama mesopotamica*), red deer (*Cervus elaphus*) and aurochs (*Bos primigenius*). Whereas small game are also abundant in the earlier cultural layers at Hayonim Cave, they clearly did not reach the high proportions characteristic of the Natufian period.

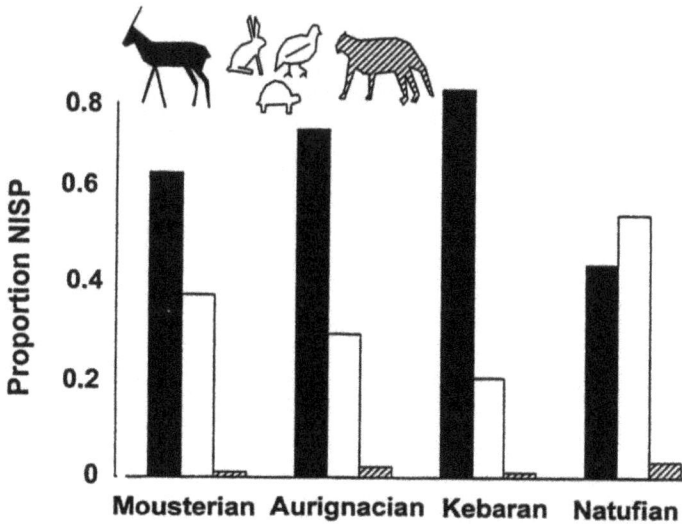

Figure 5.2. Relative proportions (NISP) of major taxonomic groups (ungulates, small game and carnivores) through time at Hayonim Cave (see Stiner et al., n.d., for NISP counts).

Even more striking than the increase in the proportion of the small game species is the change in the taxonomic composition of the small game taxa. During the Natufian period, the small game fraction is composed of relatively equal proportions of three small animal groups (see Figure 5.3). These are the slow-moving reptile species (37%), which are primarily tortoise (*Testudo graeca*) but also include two lizards (*Ophisaurus apodus* and *Agama stellio*); the avian taxa (32%), which are dominated by chukar partridge (*Alectoris chukar*) but also include other species such as rock doves (*Columba livia*) and raptors; and the small mammal group (30%), which is almost entirely composed of hare but also includes some squirrel (*Sciurus anomalus*) and hedgehog (*Erinaceus europaeus*) remains. In

sum, while several species are represented in the small game fraction, only three species, the tortoise, hare, and partridge account for more than 85% of the small animal total.

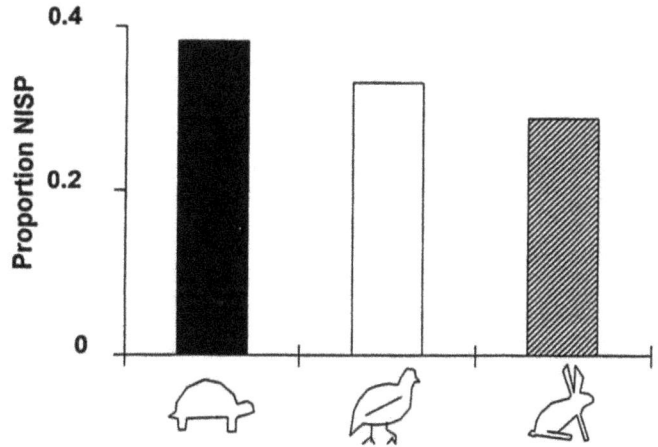

Figure 5.3. Relative proportions (NISP) of major small game taxa during the Natufian period at Hayonim Cave.

In contrast to the Natufian fauna, Mousterian small game assemblages are dominated by slow-moving species with low population turnover, particularly tortoises (see Figure 5.4). Though tortoises are a staple throughout the Hayonim sequence, they gradually decline in frequency in comparison to the fast reproducing species that are present in later periods. Moreover, a wide array of bird taxa first appear in abundance during the Aurignacian period in the early Upper Paleolithic, and then increase even further in abundance during the Natufian period. Hare, also a fast reproducer, appears in large numbers only during the Natufian period. Finally, although tortoises persevere into the Natufian period, their proportions are significantly reduced relative to the rapidly increasing frequencies of partridge and hare.

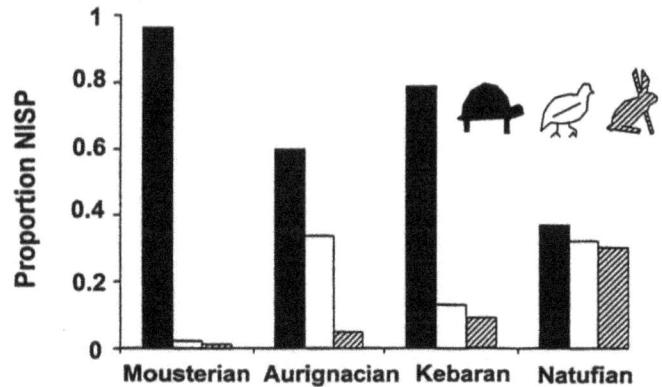

Figure 5.4. Relative proportions (NISP) of major small game taxa (reptiles, birds and small mammals) through time at Hayonim Cave (see Stiner et al. n.d. for NISP counts).

Discussion
At Hayonim Cave during the Natufian period, humans procured significantly more small animals, and, in particular, much greater proportions of partridge and hare than in earlier periods. It seems, then, that the change we see in the Hayonim Cave faunal sequence just prior to the

agricultural revolution does not reflect a broadening of the resource spectrum, but does represent a change in the relative proportion of slow, high-ranked animals in relation to fast, low-ranked ones. The substantial small game component in the Hayonim Natufian assemblage exhibits remarkable evenness across the three categories of small game taxa (reptiles, birds and small mammals). Two of these groups (birds and small mammals) are characterized by high population turnover rates and quick escape mechanisms. Without special tools, procurement of small, agile animals by humans would have required increased energy investment and reduced returns as compared to tortoise and large animals; yet as rapid reproducers, these same small species would have been more resistant to higher intensities of human exploitation. Changes in the proportions of high versus low-ranked animals clearly mirror shifts in human adaptive strategies, shifts that are arguably linked to increased population density at the regional scale. It is asserted that the sequence in which small game taxa were introduced to Paleolithic diets indicates evolutionarily significant changes in hominid foraging strategies.

Overall, the results from the Natufian period at Hayonim Cave are consistent with predictions from optimality theory and Surovell's models for increased human occupation duration at a residential site. The simulations predict that as sites are occupied for longer periods, the proportions of fast-reproducing game in the ecosystem will increase in relation to slow-reproducing game. According to optimization principles, humans are expected to shift from high to low-ranked species as local environments become depleted due to sustained use. In both cases it was predicted that the proportions of taxa such as hares and birds will increase in frequency in relation to tortoises and ungulates as the duration of site occupation lengthens. This is the pattern found in the fauna from the Natufian period at Hayonim Cave.

In light of these results, the following scenario is proposed as a reconstruction of game use at Hayonim Cave during the Natufian period. As humans occupied Hayonim Cave for increasingly longer consecutive periods, they were forced to reap large amounts of resources from the local ecosystem each year. Sustained harvesting eventually lead to the depletion of high-ranked species with low-turnover rates, namely gazelle and tortoises. The depletion of high-ranked resources in the ecosystem surrounding Hayonim Cave forced the inhabitants to include lower-ranked resources such as hares and partridges in their diets. The addition of these resources was made feasible through the use of new technologies such as nets, traps and snares. When lower-ranked species were first added to Natufian diets, their density in the environment was high; at the same time populations of the higher-ranked species were being depleted. Following their inclusion in the diet, low-ranked species were much better able than high-ranked species to maintain their population densities and to recover quickly in the face of heavy hunting disturbance. As a result, low-ranked species continued to thrive at high population densities and outnumbered high-ranked species in the local ecosystem. The availability of animals in the surrounding environment determined the encounter rates between human hunters and their prey. Greater frequencies of hare and partridge in the ecosystem, thus translated into larger proportions of hare and partridge bones in the archaeological record.

Obviously, it is impossible to pinpoint the number of days that Hayonim Cave was occupied each year. The patterns in game use at the site may have been produced by an increase in occupation duration, or in the number of people living at the site or both. Regardless of the specific details of the demographic change, there is no question that people were using the local ecosystem significantly more intensively than they were in earlier periods. The strength of the signal may be blurred by time averaging. Low-ranked species are not expected to exceed high-ranked species in the human diet until the ecosystem has been significantly impacted by human activity. This means that in environments largely untouched by human hunting, the ratio of high- to low-ranked species captured by humans will be greater at the beginning of the human occupation than the end. The proportion of high to low-ranked species changes over the course of site occupation and these proportions are averaged in the archaeological record. As a result taxa such as tortoises are not expected to disappear in the archaeological record; their proportions are only expected to decline in relation to lower-ranked prey. Time averaging only serves to weaken the signal of intense human occupation, thus if the signal is strong, the pattern gains further credibility.

Interestingly, the changes in game use at Hayonim Cave do not correlate with the climatic shifts that ranged between extremes in temperature and moisture over the course of human occupation at the site (Baruch and Bottema 1991; Bottema and van Zeist 1981). Hare, partridge, and tortoise are at least minimally represented in each of the archaeological layers, indicating that they were all locally available throughout the long occupation sequence at Hayonim Cave (see also Tchernov 1992). Finally, modern hare, partridge and tortoise populations are distributed over broad geographic areas and have generalized dietary requirements, making them well adapted to periods of climatic instability.

It is not expected that all Natufian sites will have similar proportions of animal taxa in their assemblages. Though most debate surrounding the Natufian period focuses on large residential base camps (e.g., Ain Mallaha, El Wad, Hayonim Cave and Terrace) and their associated architecture and material culture, large sites are relatively few. There is a clear dichotomy in the size of Natufian sites. In general, the large sites have been interpreted as semi-sedentary residential base camps with long periods of occupation each year (e.g., Bar-Yosef 1991; Bar-Yosef and Belfer-Cohen 1989; Goring-Morris 1995; Perrot 1966; Valla 1991). Most Natufian sites, however, are much smaller. The small sites are more commonly interpreted as seasonal resource extraction sites, and in some cases as short-term residential camps (e.g., Betts 1991; Copeland 1991; Goring-Morris and Bar-Yosef 1987; Ronen and Lechevallier 1991). It is expected that small resource extraction sites will exhibit very different signatures than the large base camps (i.e., more high-ranked species in comparison to low-ranked species).

Unfortunately, this study is confined to a single site on the archaeological landscape. The technique, however has great relevance for exploring demographic patterns on a much larger scale. This method can be used to reconstruct patterns in human demography on a regional scale to assess shifts in human population density at the Pleistocene/Holocene boundary, and, by extension, to evaluate the timing of changes during the Neolithic revolution. A reconstruction of the distribution of human populations across Southwest Asia during the Natufian period will have tremendous power to address questions of demographic imbalance that have been critical in current explanations for the agricultural transition (e.g., Bar-Yosef and Belfer-Cohen 1989, 1991; Binford 1968; Cohen 1977; Flannery 1969; Henry 1989). Finally, this technique can also be applied to a vast range of archaeological questions beyond the current example. Transitions from foraging to farming and from nomadism to sedentism can be investigated in any relevant time period worldwide by assessing patterns in game use. Overall, this study demonstrates that small game are critical for our understanding of demographic questions because they are diverse in their responses to predator pressure, and because their biological properties set distinct limits on how they can be used by humans.

Acknowledgements
Support for this research was provided by a grant from the National Science Foundation to Mary C. Stiner (SBR-9511894), the Levi Sala Care Foundation, and a Social Sciences and Humanities Research Council of Canada (SSHRC) fellowship. Mary Stiner and Todd Survoell were instrumental in providing feedback at all stages in this study. Thanks to Ofer Bar-Yosef for giving me the opportunity to participate in the Hayonim Cave Archaeological Project and to Eitan Tchernov for allowing me to study the fauna at his lab in the Department of Evolution, Systematics and Ecology at the Hebrew University in Jerusalem. Anna Belfer-Cohen provided me with a much appreciated interpretation of the complicated Natufian stratigraphy at Hayonim Cave. Thanks also to Kate Sarther, Mary C. Stiner and Todd Surovell for commenting on drafts of this paper. Finally I'd like to thank Jon Driver for organizing the symposium at which this paper was originally presented.

References

Andrews, P., 1990. *Owls, Caves and Fossils*. Chicago: University of Chicago.

Bailey, R. C. and R. J. Aunger, 1989. Net hunters vs. archers: variation in women's subsistence strategies in the Ituri forest. *Human Ecology* 17(3), 273-297.

Baruch, U. and S. Bottema, 1991. Palynological evidence for climatic change in the Levant ca. 17,000-9,000 B.P. In *The Natufian Culture in the Levant*, ed. O. Bar-Yosef and F.R. Valla. Ann Arbor: International Monographs in Prehistory, pp. 11-20.

Bar-Yosef, O., 1991. The archaeology of the Natufian layer at Hayonim Cave. In *The Natufian Culture in the Levant*, ed. O. Bar-Yosef and F.R. Valla. Ann Arbor: International Monographs in Prehistory, pp. 81-92.

Bar-Yosef, O. and A. Belfer-Cohen, 1989. The origins of sedentism and farming communities in the Levant. *Journal of World Prehistory* 3(4), 447-498.

Bar-Yosef, O. and A. Belfer-Cohen, 1991. From sedentary hunter-gatherers to territorial farmers in the Levant. In *Between Bands and States*, ed. S.A. Gregg. Center for Archaeological Investigations, Occasional Paper No. 9. Carbondale: Southern Illinois University.

Bar-Yosef, O. and N. Goren, 1973. Natufian remains in Hayonim Cave. *Paléorient* 1: 49-68.

Belfer-Cohen, A., 1988. The Natufian Settlement at Hayonim Cave. Unpublished Ph.D. Dissertation. Hebrew University, Jerusalem, Israel.

Belfer-Cohen, A., 1991. The Natufian in the Levant. *Annual Review of Anthropology* 20, 291-307.

Betts, A., 1991. The late Epipaleolithic in the Black Desert, eastern Jordan. In *The Natufian Culture in the Levant*, ed. O. Bar-Yosef and F.R. Valla. Ann Arbor: International Monographs in Prehistory, pp. 217-234.

Binford L.R., 1968. Post-Pleistocene adaptations. In *New Perspectives in Archaeology*, ed. S. Binford and L.R. Binford. Chicago: Aldine, pp. 313-341.

Boserup, E., 1965. *The Conditions of Agricultural Growth*. Chicago: Aldine.

Bottema, S. and W. van Zeist, 1981. Palynological evidence for the climatic history of the Near East 50,000-6,000 BP. In *Préhistoire du Levant: chronologie et organisation de l'espace depuis les origines jusqu'au VIe Millenaire*, ed. J. Cauvin and P. Sanlaville. Paris: Editions du CNRS, Colloques Internationaux du CNRS No 598, pp. 111-132.

Broughton, J. M., 1994. Declines in mammalian foraging efficiency during the late Holocene, San Francisco Bay, California. *Journal of Anthropological Archaeology* 13, 371-401.

Bousman, C. B., 1993. Hunter-gatherer adaptations, economic risk and tool design. *Lithic Technology* 18, 59-86.

Byrd, B.F., 1989. The Natufian: settlement variability and economic adaptations in the Levant at the end of the Pleistocene. *Journal of World Prehistory* 3(2), 159-197.

Campana, D. V. and P. J. Crabtree, 1990. Communal hunting in the Natufian of the southern Levant: the social and economic implications. *Journal of Mediterranean Archaeology* 3(2), 223-243.

Cohen, M.N., 1977. *The Food Crisis in Prehistory: Overpopulation and the Origins of Agriculture*. New Haven: Yale University Press.

Colledge, S.M., 1991. Investigations of plant remains preserved in Epipaleolithic sites in the Near East. In *The Natufian Culture in the Levant*, ed. O. Bar-Yosef and F.R. Valla. Ann Arbor: International Monographs in Prehistory, pp. 391-398.

Cope, C.R., 1991a. The Evolution of Natufian Megafaunal Communities. Unpublished Ph.D. Dissertation. Hebrew University, Jerusalem, Israel.

Cope, C.R., 1991b. Gazelle hunting strategies in the southern Levant. In *The Natufian Culture in the Levant*, ed. O. Bar-Yosef and F.R. Valla. Ann Arbor: International Monographs in Prehistory, pp. 341-358.

Copeland, L., 1991. Natufian sites in Lebanon. In *The Natufian Culture in the Levant*, ed. by O. Bar-Yosef and F.R. Valla. Ann Arbor: International Monographs in Prehistory, pp. 27-42.

Davis, S.J., 1978. The Large Mammals of the Upper Pleistocene-Holocene in Israel. Unpublished Ph.D. dissertation. Hebrew University, Jerusalem, Israel.

Davis, S.J., 1983. The age profiles of gazelles predated by ancient man in Israel: possible evidence for a shift from seasonality to sedentism in the Natufian. *Paléorient* 9:55-62.

Davis, S.J., 1989 Why did prehistoric people domesticate food animals? The bones from Hatoula 1980-86. In *Investigations in South Levantine Prehistory, Préhistoire du Sud-Levant*, ed. O. Bar-Yosef and B. Vandermeersch. Oxford: BAR International Series 497.

Davis, S.J., O. Lernau and J. Pichon, 1994. The animal remains: new light on the origin of animal husbandry. In *Le Gisement de Hatoula en Judée Occidentale, Israël*, ed. M. Lechevallier and A. Ronen. Jerusalem: Centre de Recherche Français de Jerusalem Vol. 8.

Edwards, P. C., 1989. Revising the broad spectrum revolution and its role in the origins of southeast Asian food production. *Antiquity* 63, 225-246.

Emlen, J. M., 1966. The role of time and energy in food preference. *American Naturalist* 100, 611-617.

Flannery, K.V., 1969. Origins and ecological effects of early domestication in Iran and the Near East. In *The Domestication and Exploitation of Plants and Animals*, ed. P. Ucko and G. Dimbleby. Chicago: Aldine, pp. 73-100.

Gebauer, A.B. and T.D. Price, 1992. Foragers to farmers: an introduction. In *Transitions to Agriculture in Prehistory*, ed. by A.B. Gebauer and T.D. Price. Monographs in World Archeology No. 4. Madison: Prehistory Press, pp. 1-10.

Goring-Morris, N., 1995. The Early Natufian occupation at El Wad, Mt. Carmel reconsidered. In *Nature et Culture, colloque de Liège*, ed. M. Otte. Liège: E.R.A.U.L., pp.415-425.

Goring-Morris, N. and O. Bar-Yosef, 1987. A Late Natufian campsite from the western Negev, Israel. *Paléorient* 13(1), 107-112.

Grayson, D., 1984. *Quantitative Zooarchaeology*. New York: Academic Press.

Hayden, B., 1981. Research and development in the Stone Age: technological transitions among hunter-gatherers. *Current Anthropology* 22:519-548.

Henry, D. O., 1989. *From Foraging to Agriculture*. Philadelphia: University of Pennsylvania Press.

Hillman, G.C., 1996. Late Pleistocene changes in wild plant foods available to hunter-gatherers of the northern Fertile Crescent: possible preludes to cereal cultivation. In *The Origins and Spread of Agriculture and Pastoralism in Eurasia*, ed. by D. Harris. Washington: Smithsonian Institution Press, Washington.

Hillman, G.C., S.M. Colledge and D.R. Harris, 1989. Plant food economy during the Epipalaeolithic period at Tell Abu Hureyra, Syria: dietary diversity, seasonality, and modes of exploitation. In *Foraging and Farming: the Evolution of Plant Exploitation*, ed. D.R. Harris and G.C. Hillman. London: Unwin-Hyman, pp. 240-268.

Hopf, M. and O. Bar-Yosef, 1991. Plant remains from Hayonim Cave, Western Galilee. *Paléorient* 13(1): 117-120.

Horwitz, L.K., 1996. The impact of animal domestication on species richness: a pilot study from the Neolithic of the southern Levant. *Archaeozoologia* 8:53-70.

Keeley, L.H., 1995. Protoagricultural practices among hunter-gatherers a cross-cultural survey. In *Last Hunters First Farmers*, ed. T.D. Price and A.B. Gebauer. Santa Fe: School of American Research Press, pp. 243-272.

Kelly, R., 1995. *The Foraging Spectrum: Diversity in Hunter-Gatherer Lifeways*. Washington: Smithsonian Institution Press.

Kislev, M.E., D. Nadel and I. Carmi, 1992. Epipaleolithic (19,000 BP) cereal and fruit diet at Ohalo II, Sea of Galilee. Israel. *Review of Palaeobotany and Palynology* 73, 161-166.

Krebs, J.R., D.W. Stephens and W.J. Sutherland, 1983. Perspectives in optimal foraging. In *Perspectives in Ornithology: Essays presented for the Centennial of the American Ornithologists Union*, ed. A.H. Brush and G.A. Clark Jr. Cambridge: Cambridge University Press, pp. 165-221.

Lieberman, D., 1991. Seasonality and gazelle hunting at Hayonim Cave: new evidence for sedentism in the Natufian. *Paléorient* 17: 47-57.

Lieberman, D., 1993. The rise and fall of seasonal mobility among hunter-gatherers: the case of the southern Levant. *Current Anthropology* 34(5), 599-631.

Lyman, L.R., 1994. *Vertebrate Taphonomy*. Cambridge: Cambridge University Press.

MacArthur, R.H. and E.R. Pianka, 1966. An optimal use of a patchy environment. *The American Naturalist* 100:603-609.

Neeley, M.P. and G.A. Clark, 1993. The human food niche in the Levant over the past 150,000 years. In *Hunting and Animal Exploitation in the Later Palaeolithic and Mesolithic of Eurasia*, ed. G.L. Peterkin, H.M. Bricker and P. Mellars. Archaeological Papers of the American Anthropological Association Number 4, pp. 221-240.

Perrot, J., 1966. Le gisement natoufien de Mallaha (Eynan), Israël. *L'Anthropologie* 70, 437-484.

Perry, G. and E.R. Pianka, 1997. Animal foraging: past, present and future. *TREE* 12(9), 360-364.

Pichon, J., 1984. L'Avifaune natoufienne du Levant. Thèse Dactylographiée. Université de Paris VI, Paris, France.

Pichon, J., 1991. Les oiseaux au Natoufien, avifaune et sédentarité. In *The Natufian Culture in the Levant*, ed. O. Bar-Yosef and F.R. Valla. Ann Arbor:

International Monographs in Prehistory, pp. 371-380.

Price, T.D. and A.B. Gebauer, 1995. New perspectives on the transition to agriculture. In *Last Hunters First Farmers New Perspectives on the Transition to Agriculture*, ed. by T.D. Price and A.B. Gebauer. Santa Fe: School of American Research Press, pp. 3-19.

Rabinovich, R., 1998. Patterns of Animal Exploitation and Subsistence in Israel during the Upper Paleolithic and the Epipaleolithic (40.000 - 12,500 BP), as Based upon Selected Case Studies. Unpublished Ph.D. dissertation, Hebrew University, Jerusalem.

Ronen, A. and M. Lechevallier, 1991. The Natufian of Hatula. In *The Natufian Culture in the Levant*, ed. O. Bar-Yosef and F. Valla. Ann Arbor: International Monographs in Prehistory, pp. 149-160.

Rosenberg, M., 1998. Cheating at musical chairs, territoriality and sedentism in an evolutionary context. *Current Anthropology* 39(5): 653-681.

Schoener, T.W., 1986. A brief history of optimal foraging theory. In *Foraging Behavior*, ed. A.C. Kamil, J.R. Krebs and H.R. Pulliam. New York: Plenum Press, pp. 5-67.

Soffer, O., 1985. *The Upper Paleolithic of the Central Russian Plain*. New York: Academic Press.

Speth, J.D. and S. L. Scott, 1989. Horticulture and large-mammal hunting: the role of resource depletion and the constraints of time and labor. In *Farmers as Hunters*, ed. S. Kent, pp. 71-79. Cambridge: Cambridge University Press, pp. 71-79.

Stephens, D.W. and J.R. Krebs, 1986. *Foraging Theory*. Princeton: Princeton University Press.

Stiner, M.C. and E. Tchernov, 1998. Pleistocene species trends at Hayonim Cave: Changes in climate versus human behavior. In *Neanderthals and Modern Humans in West Asia*, ed. O. Bar-Yosef and T. Akazawa. New York: Plenum Press.

Stiner, M.C., N.D. Munro and T.A. Surovell, n.d. The tortoise and the hare: small game use, the Broad Spectrum Revolution and human demographics. Ms. in preparation for Current Anthropology.

Stiner, M.C., N.D. Munro, T.A. Surovell, E. Tchernov and O. Bar-Yosef, 1999a. Paleolithic growth pulses evidenced by small animal exploitation. *Science* 283, 190-194.

Tchernov, E., 1984. Commensal animals and human sedentism in the Middle East. *In Animals and Archaeology*, ed. J. Clutton-Brock and C. Grigson. Oxford: BAR International Series 202, pp. 91-115.

Tchernov, E., 1991. Biological evidence for human sedentism in southwest Asia during the Natufian. In *The Natufian Culture in the Levant*, ed. by O. Bar-Yosef and F.R. Valla. Ann Arbor: International Monographs, pp. 315-340.

Tchernov, E., 1992. Biochronology, paleoecology, and dispersal events of hominids in the southern Levant. In *The Evolution and Dispersal of Modern Humans in Asia*, ed. by T. Akazawa, K. Aoki, and T. Kimura. Tokyo: Hokusen-Sha, pp. 149-188.

Tchernov, E., 1993. The impact of sedentism on animal exploitation in the southern Levant. In

Archaeozoology of the Near East, ed. H. Buitenhuis and A. Clason. Leiden: Universal Book Services.

Unger-Hamilton, R., 1989. Epipalaeolithic Palestine and the beginnings of plant cultivation: the evidence from harvesting experiments and microwear studies. *Current Anthropology* 30:88-103.

Valla, F.R., 1991. Les Natoufiens de Mallaha et l'espace. In *The Natufian Culture in the Levant*, ed. O.Bar-Yosef and F.R. Valla. Ann Arbor: International Monographs in Prehistory, pp. 111-122.

HUNTING THE BROAD SPECTRUM REVOLUTION: THE CHARACTERISATION OF EARLY NEOLITHIC ANIMAL EXPLOITATION AT QERMEZ DERE, NORTHERN MESOPOTAMIA

Keith Dobney, Mark Beech and Deborah Jaques
Environmental Archaeology Unit, Department of Biology, University of York, U.K.

Introduction

The site of Qermez Dere lies on the N.W. outskirts of the town of Tell Afar, on the southern. side of the Jebel Sinjar range, approximately 50 km W. of Mosul in northern Iraq (Figure 6.1). It is located immediately by the main road which by-passes Tell Afar en-route from Mosul to the town of Sinjar and lies just above the huge, flat expanse of the North Jezirah plain in the foothills of the eastern extension of the Jebel Sinjar range.

The site's importance can be understood in terms of its location and date. It is the earliest permanent settlement site (along with Nemrik, excavated at the same time by Professor Stefan Kozlowski) in Northern Iraq outside the mountain and piedmont zone of NE Iraq. Since settlement sites of such a date are extremely rare in the Near East, Qermez Dere is a significant site in unravelling the story of the beginnings of sedentary village life and the origins of farming.

The strange architecture of the settlement bears on the cultural and social development of early sedentary communities, and their exploitation of local resources by hunting and gathering gives a very important date after which cultivation and herding began. A detailed analysis of the vertebrate remains from Qermez Dere was undertaken in order to closely define the economic basis of this sedentary early neolithic hunter-gatherer community and illuminate subsistence patterns of pre/proto-domestication communities in Northern Mesopotamia.

The analysis aimed to:

- characterise the deposits over the site and attempt to identify patterns of activity and refuse disposal;
- determine and quantify the taxa represented;
- test for seasonal variation of exploitation;
- provide clues as to the immediate environs of the site and the range of exploitation;
- establish how this assemblage fits into the wider picture of late Epipalaeolithic/early neolithic animal exploitation in the Near East particularly in regard to the 'broad-spectrum' concept; and
- provide the basis for a model of proto-domestication resource exploitation in northern Mesopotamia.

The excavation

The excavation was undertaken at the suggestion of the Department of Antiquities and Heritage as the site was seriously threatened by road-building, by pipeline and communication cable trenches and by the mechanised digging of soil for making gardens around new houses of the expanding town. Early in 1986 the Directorate-General of

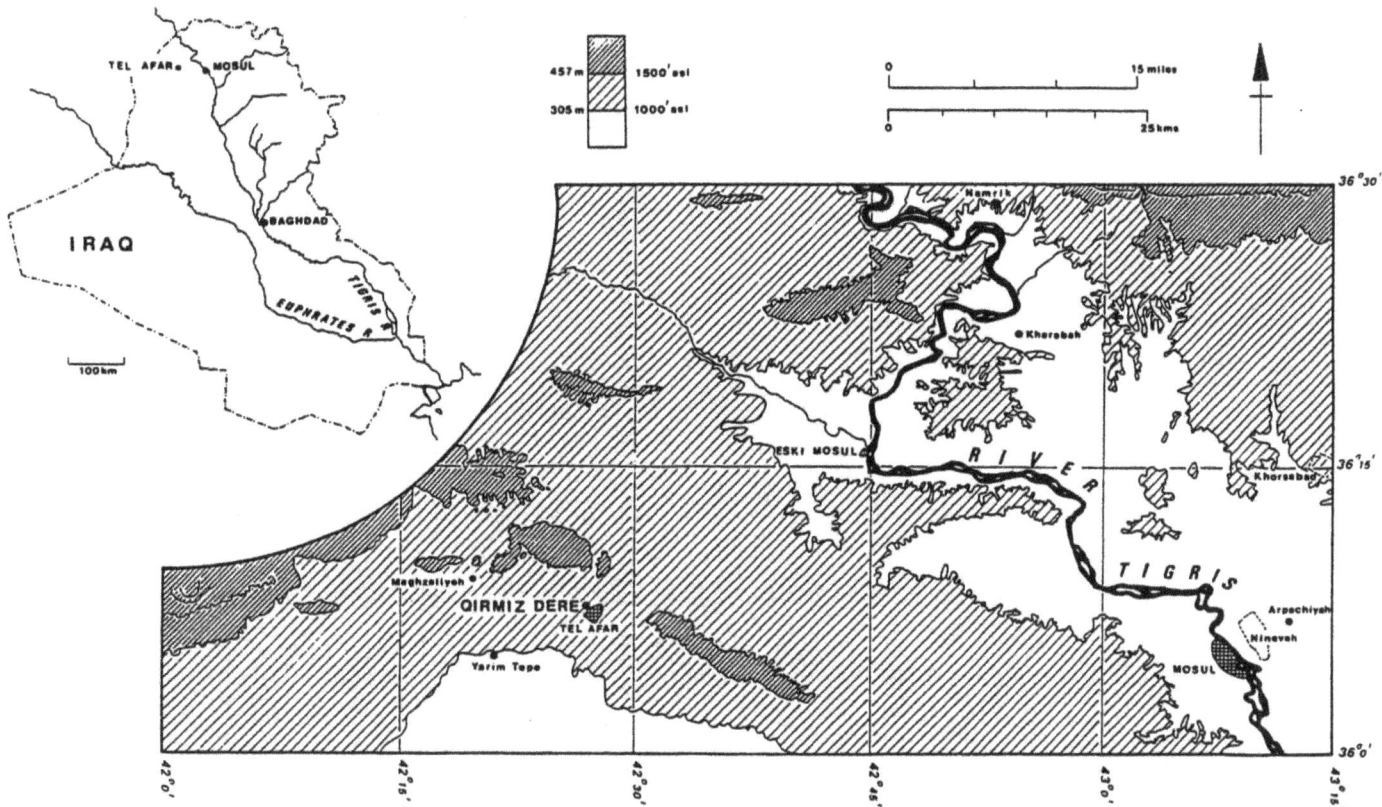

Figure 6.1. Position of Qermez Dere

47

Antiquities approached the British Archaeological Expedition in Iraq to see if they were interested in mounting an emergency rescue excavation. That October, a small trial excavation was carried out under the supervision of Dr. Ellen McAdam, and a first season of excavation led by Dr Trevor Watkins of Edinburgh University followed from mid-April to mid-May 1987 (Watkins and Baird 1987; Watkins, Baird and Betts 1989). A second season of excavation took place at Qermez Dere from early-April to mid-May 1989 (Watkins et al. 1991). The third and final season of excavation took place between early-April to mid-May 1990 (Watkins et al. 1995).

Dating

In 1992 a short series of six radiocarbon dates was undertaken on samples from Qermez Dere at the Oxford Accelerator Mass Spectrometer Laboratory (Watkins et al 1995). These revealed that the occupation of the site began some time before 8000 BC, continuing into the first half of the eighth millennium BC. With the exception of a single early date, the remaining five dates form a cohesive group, whose pooled mean lies between 8000 and 7900 BC. This series of dates accords with those radiocarbon dates from the culturally contemporary site of Tell Mureybit (N.Syria) in its late Phase 1 and Phase 2 stages. This small village settlement at Qermez Dere dates therefore to the proto-neolithic or very early aceramic neolithic, a period which witnessed the beginnings of the transition from hunter-gathering to the domestication of selected plants and animals.

Table 6.1 Total number of identified fragments

Taxon	Common name	Phase							Total
		2	2.5	3	4	5	6	7	
cf. *Pernis apivorus*	?honey buzzard	0	0	1	0	0	0	0	1
Neophron percnopterus	eqyptian vulture	0	0	0	0	0	5	1	6
Buteo spp	buzzard	0	0	0	0	0	0	1	1
Buteo buteo	common buzzard	0	1	1	1	0	1	1	5
cf. *Buteo rufinus*	?long-legged buzzard	0	1	1	0	1	8	1	12
Aquila rapax	steppe eagle	0	0	0	0	1	0	8	9
cf. *Alectoris chukar*	?rock partridge	0	0	0	0	0	1	0	1
Alectoris chukar	chukar partridge	0	0	3	0	0	0	0	3
cf. *Ammoperdix griseogularis*	?see-see partridge	0	0	0	1	0	3	1	5
cf. *Francolinus francolinus*	?black francolin	0	0	0	0	0	1	0	1
Francolinus francolinus	black francolin	0	0	0	0	1	0	0	1
cf. *Anthropoides virgo*	?demoiselle crane	0	0	0	0	0	1	0	1
Chlamydotis undulata	houbara bustard	0	0	2	0	2	16	6	26
Otis tarda	great bustard	0	0	0	0	0	1	1	2
Pterocles sp	sand grouse	1	1	5	3	5	37	5	57
cf. *Pterocles alchata*	?pin-tailed sand grouse	3	5	24	41	18	95	39	225
cf. *Pterocles orientalis*	?black-bellied sand grouse	1	0	2	8	6	7	0	24
Bubo bubo	eagle owl	0	0	0	2	0	0	0	2
Corvus corone/frugilegus	carrion crow/rook	0	0	1	0	0	2	0	3
Alaudidae	lark	4	1	4	0	0	2	0	11
	Total Bird	9	9	44	56	34	180	64	396
Hemiechinus auritus	long-eared hedgehog	*	*	*	*		*		
Meriones sp	jird			*			*		
Tatera indica	Indian gerbil		*		*				
Lepus capensis	cape hare	19	16	111	97	76	152	81	552
Vulpes vulpes	red fox	18	20	162	913	67	372	125	1677
Meles meles	badger	0	0	0	0	1	3	2	6
Felis silvestris	wild cat	1	3	2	11	3	12	11	43
Equus hemionus	onager	0	0	0	0	1	1	1	3
cf. *Equus hemionus*	?onager	0	0	0	0	0	0	1	1
Bos primigenius	wild cattle	0	0	3	1	2	3	3	12
Gazella cf. *subgutturosa*	?goitred gazelle	53	33	205	72	174	462	175	1174
Ovis orientalis	wild sheep	4	3	13	2	1	11	8	42
Caprovid	sheep/goat	2	0	9	9	10	4	4	38
Ovis/Capra/Gazella	sheep/goat/gazelle	12	10	58	20	78	323	138	639
	Total Mammal	109	85	563	1125	413	1343	549	4187

Recovery and phasing

The vertebrate assemblage from this small site was carefully controlled for context, and care was taken with quantifying the sampling process. Dry-sieving of all sediment matrix was undertaken through screens with 4mm apertures, whilst standard wet-sieve samples (60 litres maximum where context size allowed) were processed on-site through 1mm mesh. Washovers or flots from these samples were kept for analysis of palaeobotanical remains, whilst residues were further wet-sieved to 3mm (residues from 10 litre voucher samples sieved to 1mm were also retained). The combination of tight stratigraphic and contextual control, rigorous collection and quantified sampling make the recovered material unique for the region.

The stratigraphic phases of the site are as follows: Phases 0 and 1 represent superficial and disturbed topsoil deposits, whilst phases 2, 2.5 and 3 are deposits from subsequent phases of house construction (houses RAA, RAD and RAB respectively). Phases 4-6 represent the 'southern midden deposits' and phase 7 'basal soil' deposits.

Evaluation of site formation processes

A range of semi quantitative data was recorded for the bone assemblages from each context. Preservation and colour of the entire vertebrate assemblage was remarkably consistent throughout, with preservation being recorded as 'fair' and colour recorded as 'fawn' for almost all contexts. The vast majority of the Qermez Dere animal bones are homogeneous in terms of fragmentation, with most contexts containing bone fragments of between 0-2 cm. The only outliers from the dry-sieved fraction included contexts 104, RDN and RDO where more than 50% of the fragments were between 2-5 cm in length. Burning was evident on much of the material,

with low frequencies in the earliest and latest phases of the midden. This may indicate a shift in the disposal practices of the inhabitants of the site during phase 5, where higher proportions of burnt bone were noted. Later house fills also appear to contain more burnt bone, and may be explained by the utilisation of abandoned dwellings as convenient dumping areas for cooked domestic waste.

Skeletal element representation

Analysis of the representation of different skeletal elements showed that gazelle heads were poorly represented in most phases, whilst the major long bones and feet were all present. This is not the case for fox and hare remains and may reflect the fact that gazelle heads were removed elsewhere and not brought back to the site, or that the removal of horncores from the heads was carried out on a different part of the site and the heads disposed of separately to the consumption waste. It is also clear that there are striking similarities in the pattern of element representation in material from phases 4-7, representing the basal level and southern midden deposits. This evidence appears to indicate that little difference in the utilisation and disposal of gazelle remains occurred at the site throughout the sequence of occupation represented by the midden. This conclusion is corroborated by both fox, and (to some degree) by the hare remains where the pattern for phases 4-7 remains remarkably similar. For all species, major meat bearing elements, as well as distal limb bones, are represented, and it would appear that both primary butchery waste and domestic refuse were equally well represented. Although skinning marks on some of the fox, hare, and even cat bones were recorded (see below), no bias towards feet and head elements exist to suggest the primary importance of pelts over meat.

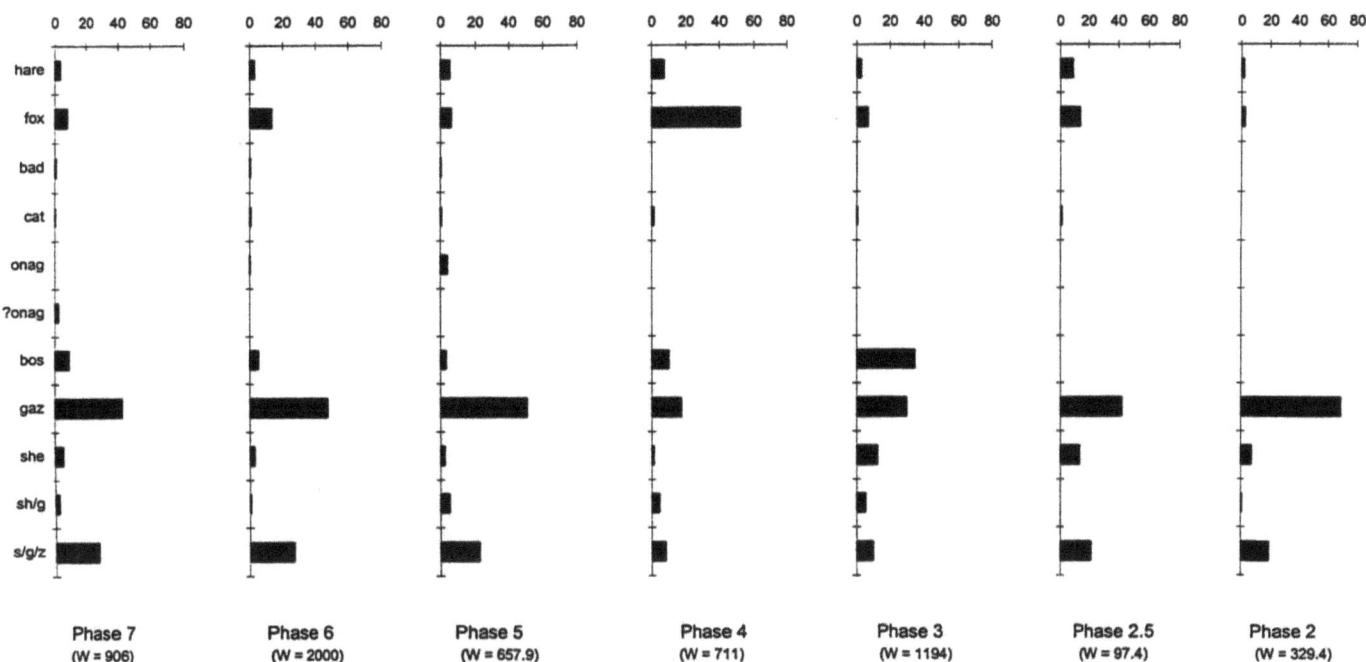

Figure 6.2. Percentage frequency of combined weight (gm) of major mammal species from wet and dry-sieved samples. (bad = badger, onag = onager, bos = *Bos primigenius*, gaz = gazelle, she = sheep, sh/g = caprovid, s/g/z = caprovid/gazelle, w=weight gm).

Occurrence and relative importance of taxa

The vertebrate assemblage recovered from excavations at Qermez Dere totalled 4,583 identified fragments. A wide range of both mammal and bird species have been identified, by far the most numerous being those of *Gazella*, *Vulpes* and *Lepus* for mammals, and *Pterocles* for birds (Table 1). When considering the varying frequencies of individual mammal species by total weight, the overall pattern is decidedly similar throughout all phases (Figure 6.2). However, a distinct peak of fox is represented in phase 4 deposits (most marked in the dry sieved assemblage), whilst gazelle appear most frequent in phase 2 deposits (again from the dry sieved material). The remains of wild caprovids also appear most frequent in later deposits (i.e. phases 2-3). Calculations of the minimum number of individuals (MNI) corroborate the pattern of frequency shown by the use of total identified fragment and total weight data and indicate that the patterns observed are certainly not artefacts of the different methods of calculating species frequency or a result of different recovery procedures. The overall patterns demonstrated for the vertebrate assemblage from Qermez Dere appear broadly similar throughout all phases.

Major mammal species

Gazelle (*Gazella* cf. *subgutturosa*)

Identifications of gazelle material from other sites has too often been based on assumptions related to modern day distribution patterns (Uerpmann 1987: 98) which may have been radically different in the past. Using biometrical data to separate species is also problematic since there is often an overlap in size between species. Figure 6.3 shows comparisons of tibia measurements between modern comparative gazelle species and those from northern Mesopotamian Neolithic sites of Qermez Dere, Nemrik and M'leefat. It is clear that values for both modern *Gazella gazella* and *Gazella subgutturosa* overlap to a considerable degree and values for all the early Neolithic specimens from the three sites discussed fall within the range of both species. Figures 6.4 and 6.5 show bivariate plots for distal humerus measurements which compare both the modern and archaeological data. Again, a major overlap occurs between values for *G. gazella* and *G. subgutturosa*, with values for the archaeological material falling towards the upper end of their distribution.

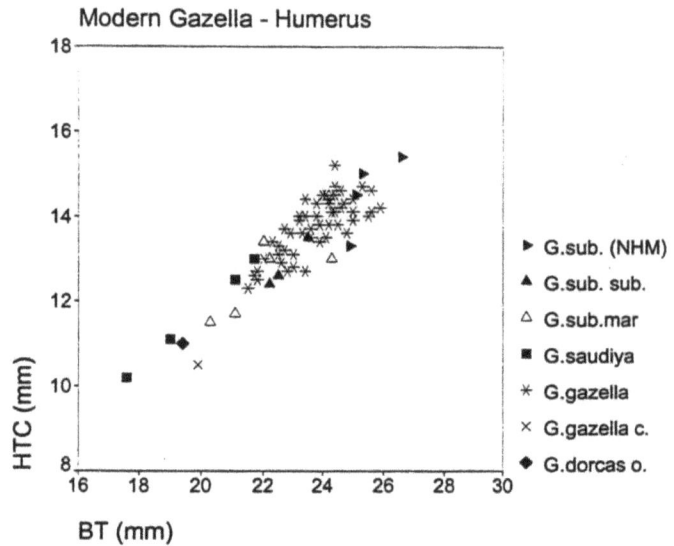

Figure 6.4. Bivariate plot of modern gazelle humeri. See Figure 6.3 for key to species.

On the basis of their modern day distribution, the goitred gazelle (*G. subgutturosa*) is the only species to occur in the lowlands and foothills north and east of the river Tigris and probably all of Mesopotamia (Uerpmann 1987: 98). *Gazella gazella* appears to be confined to east of the Euphrates, in the hills and coastal area of the southern Levant (Uerpmann 1987: 100) and there is no firm archaeological evidence of an eastern distribution beyond the site of Ksar 'Akil in central Lebanon (Hooijer 1961). The geographical location of Qermez Dere (and that of Nemrik) falls within the modern range of *Gazella subgutturosa*, and at the centre of the northern distribution of other proto and late Neolithic sites where goitred gazelle remains have been definitely identified (Uerpmann, 1982, Bibikova 1981, Turnbull and Reed 1974). On the basis of general size and distribution therefore, the gazelle remains from Qermez Dere, Nemrik and M'leefat are

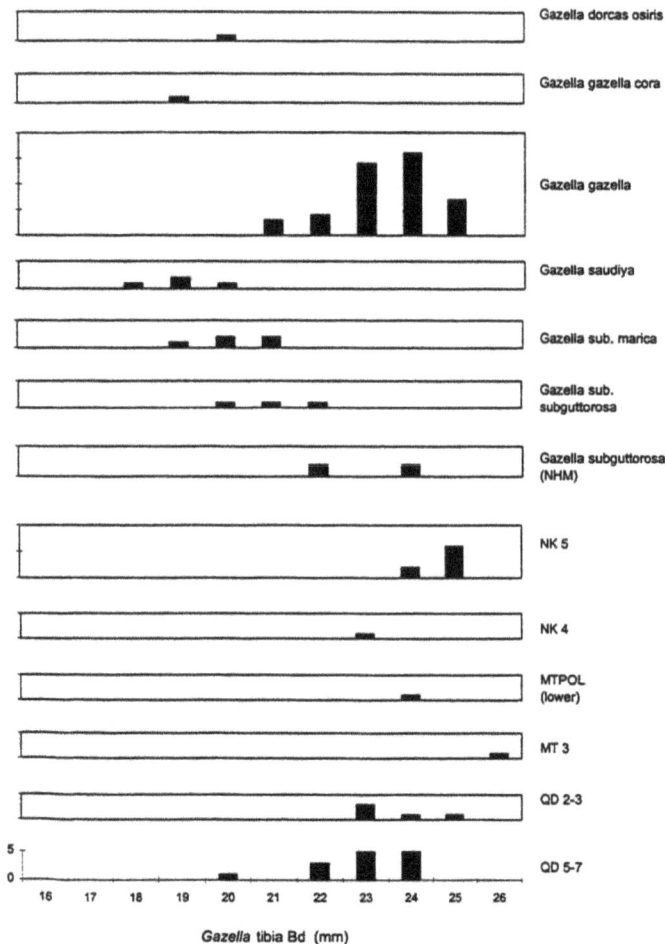

Figure 6.3. Distal breadth (Bd) of archaeological and modern gazelle tibiae. (QD = Qermez Dere; MT and MTPOL = M'lefaat; NK = Nemrik; NHM = Natural History Museum. All other data from Environmental Archaeology Unit comparative specimens).

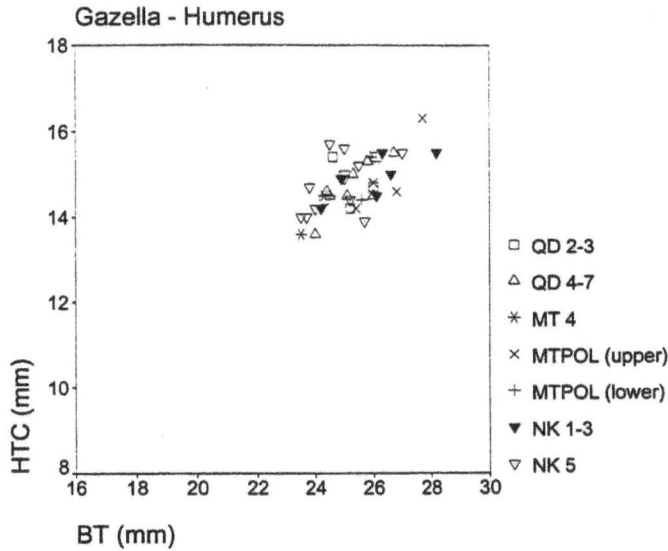

Figure 6.5. Bivariate plots of archaeological gazelle humeri. See Figure 6.3 for key to sites.

most likely that of *Gazella subgutturosa*. The few poorly preserved horncore fragments seem to support this

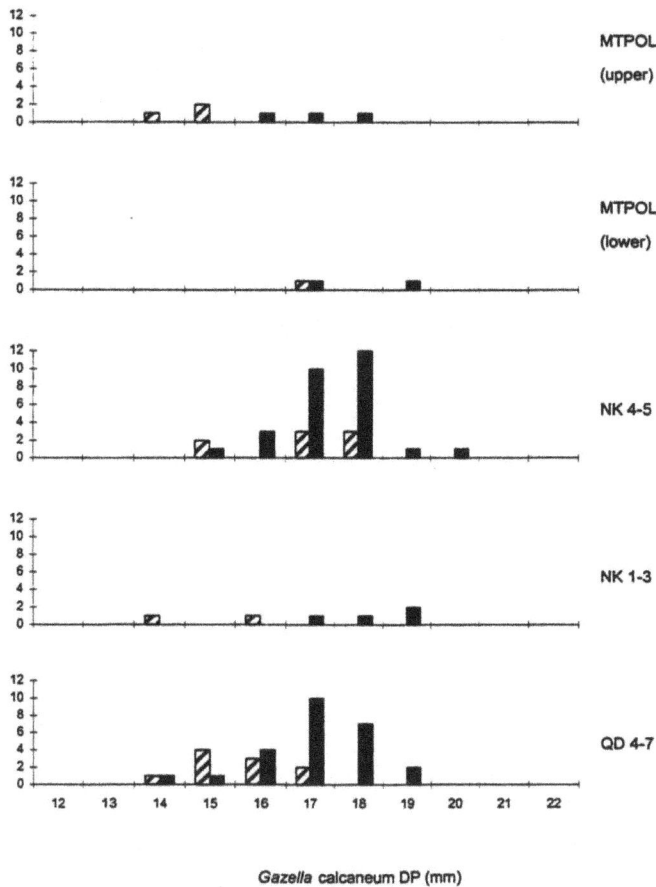

Figure 6.6. Anterior-posterior depth of archaeological gazelle calcanei (DP). Solid bars = fused proximal epiphysis. Open bars = unfused proximal epiphysis.

conclusion, although the presence of two female horncores (similar to those recovered from Douara cave [Payne 1983]) may also indicate the occurrence of the *marica* subspecies (Uerpmann 1987:101). It is certainly the case that several small tibia breadth and calcaneum length measurements, from Qermez Dere, fall within the range of those presented for modern *G. s. marica* comparative specimens.

As a result of the fragmented nature of both teeth and long bones, little information regarding the age at death profile of the gazelle population from Qermez Dere could be gathered. Where post-cranial fragments did allow the assessment of the fusion status of individual elements, the frequency of fused (and thus skeletally mature) specimens, was always high, on average around 83%. Figure 6.6 shows gazelle calcaneum DP (anterior-posterior depth of the proximal articulation at fusion line) measurements. The pattern found at Abu Hureyra, northern Syria, has been interpreted as one which implies seasonal killing of gazelles, with three separate peaks of newborn, yearlings and adults (Legge and Rowley Conwey pers comm.). At Qermez Dere, although almost no newborn gazelle bones were recovered, two possible peaks of yearlings and adults could also be tentatively interpreted as evidence of seasonal killing.

Fox (*Vulpes vulpes arabica*)
Definitive identification of the fox species at Qermez Dere is made more problematic by the fact that both the red fox (*Vulpes vulpes arabica*) and Ruppell's sand fox (*Vulpes ruppelli*) are listed by Hatt (1959) as being present within the Assyrian plains and the foothills of Iraq. Identification was exacerbated by the fragmented

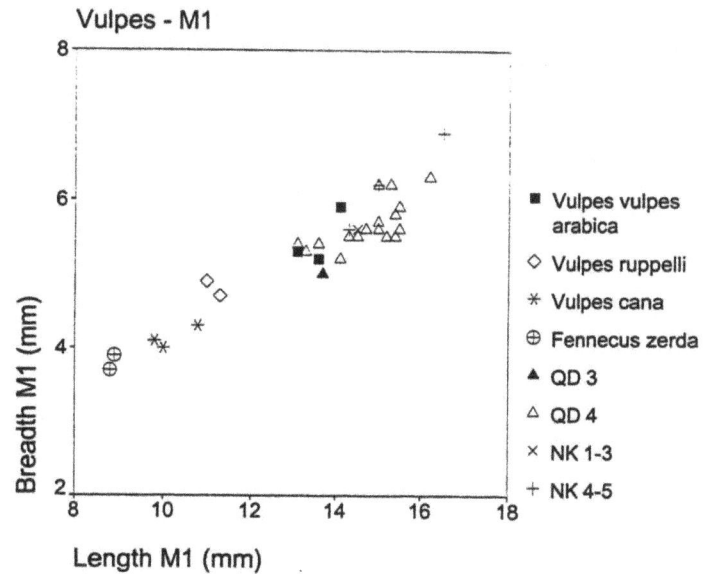

Figure 6.7. Bivariate plot of modern and archaeological fox carnassial (M1).

nature of the material, with the result that only post-cranial elements and isolated teeth were recovered. Although numbers of actual specimens are low, measurement data from isolated mandibular first molars (M1), however, clearly show that the fox remains from Qermez Dere and Nemrik most certainly represent the red fox, rather than the smaller *Vulpes ruppelli*, *Vulpes cana* or *Fennecus zerda* (Figure 6.7).

Almost all the remains represented adult individuals, with only 14 of a total of 401 fragments (where epiphyseal fusion data could be gleaned) being unfused. There is little doubt that the red fox remains from Qermez Dere represent the remains of consumption, since many of fox bones are burnt. In a few cases, evidence of small parallel cut marks attest to the probable removal of the pelt.

Hare (*Lepus capensis*)

The remains of hare were also recovered in large numbers from the Qermez Dere assemblage. As was the case for the fox remains, many of the hare bones were also burnt and several again showed evidence of skinning marks on distal limb elements. Although fragmented, a moderate biometrical archive was collected from all phases at Qermez Dere and compared to those data from several Epipalaeolithic and later Neolithic eastern Jordanian desert sites reported by Martin (1994). At these sites preliminary analysis showed that a

significant size decrease had occurred in the hares between the Epipalaeolithic and the later Neolithic, although (unfortunately) no data of early Neolithic date was available for the eastern desert region. The data set from Qermez Dere falls within this missing time period, and Figure 6.8 clearly show that the hares from the early Neolithic of Northern Mesopotamia were comparable in size to those from the Epipalaeolithic of the eastern Jordanian desert.

Tentative evidence of a general size decrease has been noted in a range of species (Davis 1977, 1981, 1982, Davis et. al.1994 & Martin 1994) during the early Holocene of the middle east, and attributed to warming of the climate. If the hares at Qermez Dere are following Bergmann's rule, then it can be inferred that the temperature gradient at early neolithic Qermez Dere was still that of the later Pleistocene and the supposed increase in temperature (as reflected through the decrease in size of foxes and hares) had yet to take effect. However, with such limited datasets, care should be taken not to oversimplify what is almost certainly a more complex picture. For example, differences in geography and terrain of the North Jezira plain and the eastern Jordanian desert may just as likely explain the size differences between the assemblages, although it is interesting that the addition of the Qermez Dere data does nothing to change the clear differences in size between the early and later eastern Jordanian datasets.

Other mammal species

Mammal species which appeared to have played a relatively minor role in the economy of the site include wild sheep (*Ovis orientalis*) and the bezoar goat (*Capra aegagrus*), the aurochs (*Bos primigenius*), the onager (*Equus hemionus*), a mustelid (possibly badger [*Meles meles*] or ratel [*Mellivora capensis*]) and the wild cat (*Felis silvestris*). Aurochs was represented at Qermez Dere in extremely small numbers, its apparent high frequencies in certain contexts being exaggerated by the fact that a single *Bos* element weighs perhaps 100 times as much as an equivalent fox bone. This comparison is not entirely without justification, since the amount of meat from a large bovid or equid would far outweigh that collected from a single hare or fox carcase. Thus, their economic importance to the inhabitants of the site should be magnified in order to account for their larger meat yield. The low frequency of cattle remains at Qermez Dere is in contrast to their apparent high frequencies in PPNB deposits at Nemrik (Lasota-Moskalewska 1994). This may be a result of the later date of the Nemrik material and the probable closer proximity of the site to upland forest cover. Although it is generally believed that aurochs in western Europe inhabited heavy forests, it is not known whether they were also forest dwellers in Mesopotamia, or whether they inhabited principally the grassy steppes or the river valleys (Hatt 1959:66). In this context, it is interesting to note that *Bos primigenius* remains were identified at the supposed late Hassuna sites of Ginnig and Khirbit Garsour which are situated well onto the North Jezira plain due south of the Qermez Dere settlement (Dobney and Jaques unpublished manuscript). Other sites in the region where aurochsen remains have been identified include Bouqras (Hooijer 1966, Clason 1977 and 1981), Tell es-Sinn (Clason 1980), Umm Dabagiyah (Bökönyi 1978), Mlefaat

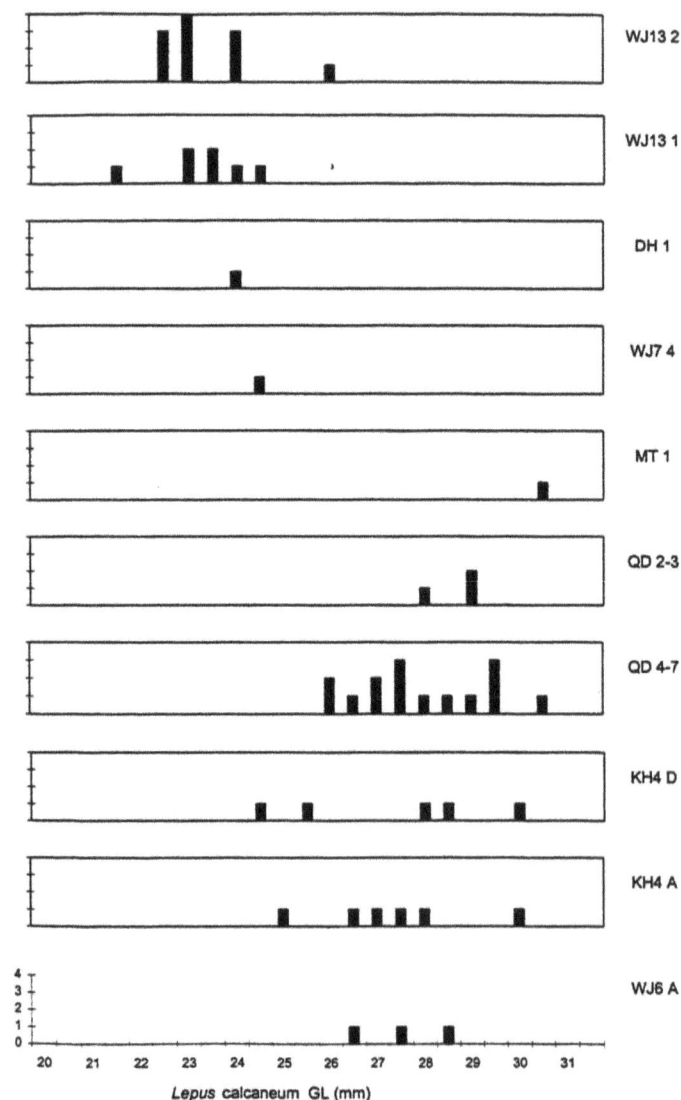

Figure 6.8. Greatest length (GL) of archaeological hare calcanei. All sites below QD4-7 are Epipalaeolithic eastern Jordanian. All sites above QD2-3 are late Neolithic eastern Jordanian.

(Turnbull 1983), Jarmo, Matarrah and Karim Shahir (Stampfli 1983), and Palegawra (Turnbull & Reed 1974).

The onager was once abundant on the plains of Iraq, however, over-hunting with the aid of vehicles and rifles has led to its probable extinction in Iraq, probably as recently as the early part of this century. One of the last herds of onager was in fact recorded near to the Jebal Sinjar (30 km North of Qermez Dere) in 1927 (Hatt 1959: 23). Onager have also been identified at Yarim Tepe (Bibikova 1980).

The scarcity of large mammal remains from the Qermez Dere assemblage is not necessarily proof that these species were not heavily exploited. If the kill sites of these animals were some distance from the settlement, it would be expected that primary butchery of the carcase would have occurred elsewhere. Distal limb elements and the head would be left, and perhaps the primary meat-bearing bones would have the meat filleted from them to reduce weight for transportation back to the site. However, in the absence of evidence, it must be assumed that large mammals were not heavily exploited.

The wild sheep *(Ovis orientalis)* must have been once more widespread in the middle east. Today they are extremely rare in the region, existing in isolated regions along the edges of mountain ranges in south-central Turkey, Armenia, Azerbeidjan down to the southeastern end of the Zagros mountains (Uerpmann 1987: 127). Sheep are generally small mountain-dwelling ruminants, preferring more open, undulating highlands than the goats.

Both cat and badger/ratel remains were recovered in small quantities from the Qermez Dere assemblage. The wild cat *(Felis silvestris)* is today distributed throughout Arabia and generally prefers rocky areas, but can also occur on flat open plains where there is available refuge, e.g. in fox-holes (Kingdon 1990: 98). The general size and morphology of the large mustelid bones indicate the presence of badger (cf. *Meles meles*) rather than the ratel or honey badger *(Mellivora capensis)*, although only two comparative specimens of Mellivora capensis were located in the Natural History Museum (one from an un-provenanced location, the other from Uganda). The fact that some of the archaeological remains of these species were burnt (some also showed sign of skinning marks) suggests that, like fox and hare, they were both eaten and utilised for their pelts.

Small mammals
Long-eared hedgehog (*Hemiechinus auritus*)
Perhaps the most numerous small mammal remains recovered from the Qermez Dere assemblage were those of hedgehog. Although particularly fragmentary, a single mandible and two maxillary fragments (with teeth still intact) were identified as *Hemiechinus auritus* (Harrison pers. comm.). The size and morphology of the teeth matched *Hemiechinus auritus* rather than the Ethiopian hedgehog (*Paraechinus aethiopicus*), which is also larger in size, or the black hedgehog (*Paraechinus hypomelas*). As was the case with the fox, hare and other smaller 'medium mammal' (MM2) remains, many of the hedgehog bones were also burnt. This almost certainly indicates that hedgehog was consumed by the Qermez Dere inhabitants.

Other small mammals identified on the basis of dental remains (Harrison and Bates pers. comm.) include the Indian gerbil (*Tatera indica*). It has been found at many localities in Iraq (Harrison 1968), with the site of Qermez Dere being located at northern edge of its present day distribution. It is a heavily built rat-like gerbil which generally inhabits agricultural land not far from water, although it can live in a wide variety of habitats ranging from semi-deserts to forests.

Jird sp. (*Meriones* sp.)
The teeth of Jirds are generally much smaller than *Tatera indica*, having cusps which are joined by 'small bridges' (Harrison pers. comm.). Identification was only possible to genus, with several species having present day distributions which include Iraq. These include Tristam's jird (*Meriones tristrami* (Thomas, 1892)), the Libyan jird (*Meriones libycus* (Lichtenstein, 1823)) and Sundevall's jird (*Meriones crassus* (Sundevall, 1842). All are robust, rat-like gerbils with relatively wide habitat preferences which render them of little interpretative value. None showed signs of burning, and are likely in fact to represent intrusive remains not associated with site occupation.

The avian fauna
The bird bones from Qermez Dere were numerous and represented a wide range of species. The recovery of so many fragments was primarily a result of the implementation of systematic wet and dry-sieving, linked with the fact that many of the bones were, although fragmented, particularly well preserved. Table 1 shows the range and numbers of identified fragments recovered from each phase.

The importance of sandgrouse (*Pterocles* sp.)
By far the most numerous remains were those of the sandgrouse (*Pterocles*). Two species are common in the region today; the pintailed sandgrouse *(Pterocles alchata)* and the black-bellied sandgrouse (*Pterocles orientalis*). Differentiating between the various sandgrouse species on the basis of skeletal morphology is extremely difficult and here it has been tentatively undertaken on the basis of the few comparative specimens held in the Natural History Museum bird collection at Tring. It was found that *P. orientalis* was somewhat more robust than *P. alchata* and the definitive identifications in the table are made on this basis alone. Biometrical data also proved somewhat inconclusive as can be seen from Figures 6.9 and 6.10.

The few comparative specimens of the different sandgrouse species (two species of sandgrouse [*P. exustus* and *P. senegallus*] not found in the region today, are also included as controls) separated rather well using bivariate plots of lengths and breadths of the major long bones. However, the archaeological material was mostly too fragmented to allow any length measurements to be taken. Width measurements of articular ends and shaft diameters were more common and these were therefore utilised.

From coracoid (Figure 6.9) and tarsometatarsus (Figure 6.10) measurements it can be seen that the sandgrouse from Qermez Dere do not overlap directly in size with with any of the modern comparative sandgrouse specimens. Those individuals represented by coracoids appear to be less robust

Coracoid

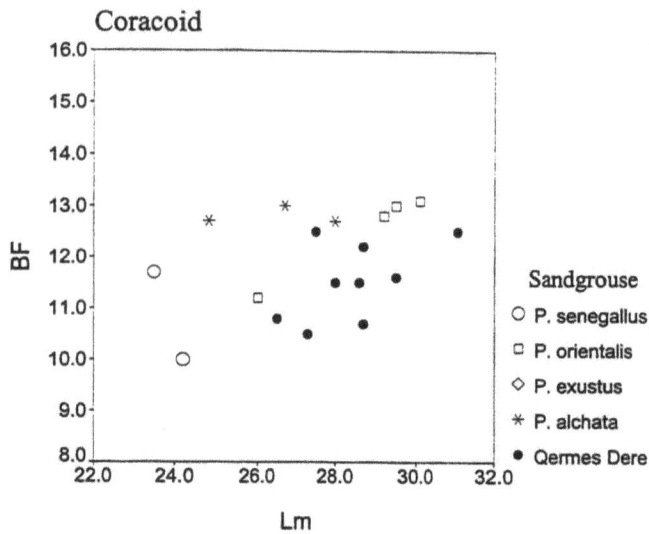

Figure 6.9. Bivariate plot of modern and archaeological sandgrouse coracoids.

than either *P. alchata* or *P. orientalis*, whereas tarsometatarsi values fall above the upper end of the *P. alchata* range. Although by no means certain, it is proposed that the vast majority of sandgrouse remains from Qermez Dere are most probably from *P. alchata*, with fewer numbers of more robust individuals probably representing *P. orientalis*.

Tarsometatarsus

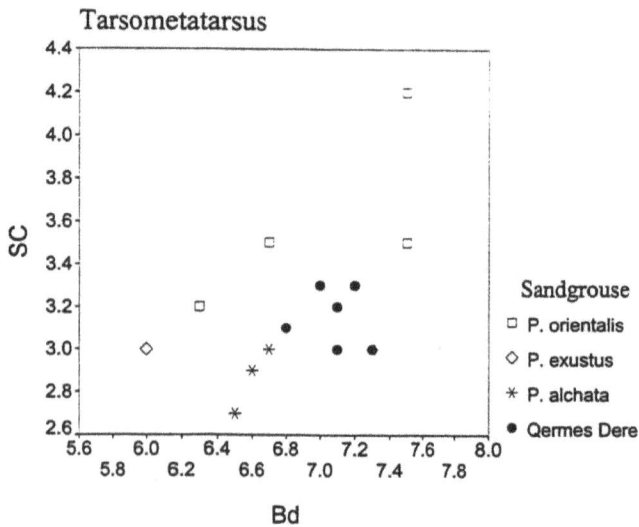

Figure 6.10. Bivariate plot of modern and archaeological sandgrouse tarsometatarsi.

The possible significance of raptors
Of particular interest are the remains of various large raptor species which appear to be represented almost exclusively by leg elements (mainly the tarsometarsus and phalanges). It is open to debate as to whether this bias in favour of leg bones in raptors is real, since when limited data for the game birds (i.e. *Pterocles*, *Ammoperdix*, *Francolinus* or *Alectoris*) are compared, similar biases are also evident. This may be a result of a combination of factors e.g. simply the ease of

identification of certain elements over others, or more complex taphonomic factors affecting disposal and preservation. However, many sites of later Epipalaeolithic and early Neolithic date from across the region show the same broad pattern for raptor remains. It is therefore probable that the presence of raptors at these sites has a different significance to those species which are thought to be primarily consumption refuse. Their presence has traditionally been interpreted as reflecting either consumption refuse or more symbolic or religious activities (Solecki and McGovern 1980). However, the data could also support another hypothesis which has, as yet, not been fully explored, i.e the capture, keeping and training of birds of prey, and the possible first faltering steps towards falconry (Dobney forthcoming). The significance of the raptor remains from proto and early Neolithic sites in the middle east does not have to reside in a single explanation. The fact that birds may have been hunted as food, or tamed, managed, and even used for hunting certainly does not preclude their importance as symbolic or totemic icons.

Placed within the cultural and environmental framework of the early Holocene, the broad spectrum sites of the middle east are extremely significant in that they represent recently sedentary human groups who still utilise hunting and gathering as an economic base. The theory of rapidly declining resources, as a result of intensive exploitation within a fixed territory, seems to be a plausible explanation for the apparent shift in focus towards smaller less rewarding species which funnelled many of these groups inextricably down the road towards agriculture. What is perhaps most significant about a falconry hypothesis at these broad-spectrum sites, is their temporal proximity to the beginnings of domestication of sheep and goat. Could experimentation with taming and management of raptors, either as a new hunting strategy and/or for religious purposes, have acted as a prelude to the beginnings of the experimentation with larger mammals? Supporting this hypothesis is the fact that the domestication of the dog took place even earlier, a species that is thought to have been primarily utilised as a hunting aid.

On the basis of all the available evidence, the significance of raptor bones recovered from numerous late Epipalaeolithic and early Neolithic sites in the middle east remains very much open to debate. Whether they simply represent domestic food refuse, symbolic artefacts or the remains of tamed and managed birds is still far from clear. The falconry hypothesis is therefore presented here merely as a possible alternative explanation, worthy of further, more critical consideration.

Vertebrate remains and palaeo-environmental reconstruction
Both the mammal and bird remains recovered from the site portray a consistent picture of the surrounding environment of the site, one not dissimilar to that which exists today. If present day preferences of the animal species identified in the Qermez Dere assemblage reflect those of 10,500 years ago, then a stony, semi-arid, open environment, possessing little in the way of tree cover, was the dominant habitat exploited. The complete absence of remains of waterfowl perhaps indicates that permanent water was not present near the site

and that the adjacent waddy contained only a seasonal waterflow. Sandgrouse only congregate in numbers once a day to drink from standing water and these may perhaps have been caught in the waddy when water was present.

The absence of wild pig (*Sus scrofa*) from Qermez Dere can perhaps be explained by their need for dense cover, shade and moist habitats, which would exclude them from more arid areas. However, it was reported to one of the authors (KD) by a long-term resident of Tel Afar, that wild boar once occurred locally in the wadi directly adjacent to the site, and that it was occasionally hunted for sport. Similarly, the absence of any cervid remains from the assemblage and large bovid also indicates that open woodland cover no longer existed in the vicinity of the site by 10,500BP. Wild pig, deer and cattle remains have been recovered from the contemporary site of Nemrik (Lasota-Moskalewska n.d.), in addition to other forest species such as beaver (*Castor fiber*) and even leopard (*Panthera pardus*). This is not wholly surprising, since the site was situated in an ecological transition zone between the Tigris floodplain and the foothills of the Taurus mountains, where a range of habitats (such as dense marshy thickets, open park woodland and denser upland forest) could be exploited.

Discussion

The vertebrate assemblage from Qermez Dere represents one of the few systematically recovered collections from Northern Mesopotamia of any date. It contributes vital data to our understanding of the economic basis of the early Neolithic in a geographical and temporal hiatus between the numerous sites of the Levant to the west and Persia to the east. The general characteristics of the Qermez Dere assemblage fits very well with all other sites of a similar period from the middle east, i.e. a reliance on the hunting of gazelles a more intensive exploitation of a range of smaller mammals (particularly fox and hare) and birds, and the absence of domestic sheep and goats. Sites exhibiting these characteristic vertebrate assemblages are supposed to reflect so-called 'broad-spectrum' economies and the phenomenon they represent has been termed the 'broad spectrum revolution' (Flannery 1969). It appears to be a widespread phenomenon across the middle east during the later Pleistocene, and is most closely linked with the emergence of the early pre-agricultural sedentary settlements dating from around 10,000 BP. Sites where broad spectrum economies have been recognised are those characterised by an apparent shift away from larger prey species towards a heavier reliance on smaller mammal species, birds, fish and even reptiles. Edwards (1989), in his re-examination of a number of Levantine Pleistocene vertebrate faunas, showed that many of these species were exploited by humans throughout much of the upper Pleistocene, and that the smaller species (such as fox and hare) were consistently exploited throughout the last 100,000 years. He concluded that no apparent increase in vertebrate diversity was visible through the Pleistocene and that the 'broad-spectrum' concept was an artefact of better recovery methods and the need to explain the beginnings of domestication.

It is certainly true, when considering the diversity of species present at individual sites, that many were indeed exploited by man during the later Pleistocene. However, when the relative frequencies of the various taxa are considered, there is no doubt that late Epipalaeolithic and early Neolithic human groups focused more intensively on gazelles and smaller taxa. Indeed, rather than a broadening of the available vertebrate prey spectrum, a shift in emphasis, or an intensification towards particular parts of the existing spectrum, occurred. This is certainly the case at Qermez Dere.

An increase in sedentism may have led to a greater degree of land ownership as proto-farming villages became increasingly dependent for their food resources within a fixed catchment area. This may well have led to territorial conflicts (Watkins 1992) of which we have limited evidence. For example, the massive walls around the site of Jericho suggest a defensive role, whilst more direct evidence was recovered from the site of Nemrik in the form of human skeletal remains with projectile points embedded in them. At Qermez Dere, numerous flint projectile points showed characteristic impact fractures to the tips, whilst the human skulls also found at this site (and a number of others) could well be enemy trophies and not revered ancestors as traditionally interpreted.

Whether demographic stress (Binford 1968, Cohen 1977), organisational changes in human society (Bender 1978, Price and Brown 1985), changes in the management of resources (Flannery 1973, Redman 1977) or environmental change (Davis 1977, 1981 & 1982) were major catalysts towards sedentism (and the associated reliance on gazelles and smaller species), it is clear that is was a widespread phenomenon throughout the middle east, and certainly present in Northern Mesopotamia during the early Neolithic. The evidence from Qermez Dere is therefore crucial in linking the different eastern and western flint tool traditions by a remarkably similar economic basis. A reappraisal of the bird of prey remains has also led to the construction of an hypothesis which suggests that early Neolithic man was possibly experimenting with falconry. This perhaps somewhat eccentric idea is presented in the hope of further critical consideration (see Dobney forthcoming).

Acknowledgements
This work was supported by grants from the Natural Environment Research Council and the British School of Archaeology in Iraq. The authors are grateful to Dr Trevor Watkins (University of Edinburgh) for providing much of the initial impetus and support as site director, and who supplied the archaeological information. Also to Simon Parfitt (Department of Palaeontology, Natural History Museum, London) who assisted with access to comparative specimens in the Museum and provided the vital spark for the falconry hypothesis, and to David Harrison and Paul Bates who provided access to the Harrison Zoological Museum comparative collections, and who kindly identified the hedgehog and gerbil bones from the assemblage.

Access to various avian comparative collections (those of the Natural History Museum at Tring, The Royal Africa Museum and the University of Ghent , Belgium) were kindly arranged by Don Smith, Dr Wim Van Neer, Dr Anton Ervynck and Prof. Achille Gautier. We are also extremely grateful to colleagues (Dr Simon Davis of

English Heritage, Dr Tony Legge of Birkbeck College, London, Dr Louise Martin of University College, London, Dr Alicja Lasota-Moskalewska, University of Warsaw who kindly provided unpublished data cited in this report).

Dr Simon Davis assisted greatly with the provision of offprints from his personal collection, whilst Professor

Stefan Kozlowski and Dr Alicja Lasota-Moskalewska of the Institute of Archaeology, University of Warsaw, provided welcome hospitality during a visit to Poland to collect comparative zooarchaeological data from the sites of Nemrik and M'lefaat.

References

Bender, B., 1978. Gatherer-hunter to farmer: a social perspective. *World Archaeology* 10, 204-222.

Bibikova, V. I., 1981. Animal husbandry in northern Mesopotamia in the 5th millennium B.C. (Materials from Halafian settlements at Yarim Tepe). In E*arliest Agricultural Settlements of Northern Mesopotamia*, ed. R.M.Munchaev and N.I. Merpert. Moscow, pp. 299-307.

Binford, L.R., 1968. Post-Pleistocene adaptations. In *New Perspectives in Archaeology*, ed. S.R. and. L.R. Binford. Aldine: Chicago, pp. 313-341.

Bökönyi, S., 1978. Environmental and cultural differences as reflected in the animal bone samples from five early Neolithic sites in southwest Asia. In *Approaches to faunal analysis in the Middle East*, ed. R.H. Meadow and M.A. Zeder. Harvard University: Peabody Museum of Archaeology and Ethnology, pp. 57-62.

Clason, A.T., 1977. Bouqras, Gomolava en Molenaarsgraaf, drie stadia in de ontwikkeling van de veeteelt. *Museologia* 7, 54-64.

Clason, A.T., 1980. The animal remains from Tell es Sinn, compared with those from Bouqras. *Anatolica* 7, 35-53.

Clason, A.T., 1981. The faunal remains of four prehistoric and early historic sites in Syria. In *Contributions to the Environmental History of Southwest Asia*, ed. W. Frey and H. P. Uerpmann. Beihefte zum Tübinger Atlas des Vorderen Orients. pp. 191-196.

Cohen, M.N., 1977. *The Food Crisis in Prehistory: Overpopulation and the Origins Of Agriculture*. New Haven and London: Yale University Press.

Davis, S. J. M., 1977. Size variation of the fox, Vulpes vulpes in the palaearctic region today, and in Israel during the late Quaternary. *Journal of Zoology* 182, 343-351.

Davis, S. J. M., 1981. The effects of temperature change and domestication on the body size of Late Pleistocene to Holocene mammals of Israel. *Paleobiology* 7, 101-114.

Davis, S .J. M., 1982. Climatic change and the advent of domestication: the succession of ruminant artiodactyls in the late Pleistocene-Holocene in the Israel region. *Paléorient* 8, 5-15.

Davis, S. J. M., Lernau, O. and Pichon, J., 1994. The animal remains: New light on the origin of animal husbandry. In *Le Gisement de Hatoula en Judée Occidentale, Israël*, ed. M. Lechevallier and A. Ronen. Mémoirs et Travaux du Centre de Recherche Français de Jerusalem. Paris: Association Paléorient.

Dobney, K., (forthcoming). Flying a kite at the end of the ice age!: new perspectives on the possible significance of raptors in Near Eastern proto-Neolithic assemblages.

Edwards, P.C., 1989. Revising the broad spectrum revolution and its role in the origins of southwest Asian food production. *Antiquity* 63, 225-46.

Flannery, K.V. 1969. Origins and ecological effects of early domestication in Iran and the Near East. In *The Domestication and Exploitation of Plants and Animals*, ed. P.J. Ucko and G.W. Dimbleby. London: Duckworth, pp. 73-100.

Flannery, K.V., 1973. The origins of agriculture. *Annual Review of Anthropology* 2, 271-310.

Hatt, R.T., 1959. *The mammals of Iraq.* Ann Arbor: Museum of Zoology, University of Michigan, Miscellaneous Publications 106.

Harrison, D.L., 1968. *The Mammals of Arabia, Vol.2 - Carnivora, Hyracoidea, Artiodactyla*. London: Ernest Benn.

Hooijer, D.A., 1961. The fossil vertebrates of Ksar 'Akil, a Palaeolithic rock shelter in the Lebanon. *Zoologische Verhandelingen* 49. Leiden.

Lasota-Moskalewska, A., (nd). Animal remains from houses 1 and 1A in Nemrik 9 (Iraq).

Martin, L.A.. 1994. Hunting and Herding in a Semi-arid Region - An Archaeozoological and Ethological Analysis of the Faunal Remains from the Epipalaeolithic and Neolithic of the Eastern Jordanian Steppe. DPhil.: University of Sheffield.

Payne, S. 1983. The animal bones from the 1974 excavations at Doura Cave. In *Paleolithic Site of the Doura Cave and Paleogeography of Palmyra Basin in Syria - Part III*, ed. K. Hanihara and T. Akazawa. Tokyo: University of Tokyo Press, pp. 1-108.

Price, T. D.and .Brown, J. A. ed., 1985. *Prehistoric Hunter-Gatherers: the Emergence of Cultural Complexity*. Academic Press: Orlando.

Redman, C. L., 1977. A model for the origins of agriculture in the Near East. In *Origins ofAgriculture*, ed. C.A. Reed. The Hague: Mouton, pp. 543-67.

Solecki, R. L.and McGovern, T. H. 1980. Predatory birds and prehistoric man. In *Theory and Practice: Essays Presented to Gene Weltfish*, ed. S. Diamond. The Hague: Mouton, pp. 79-95.

Stampfli, H. R., 1983. The fauna of Jarmo, with notes on animal bones from Matarrah, the 'Amuq and Karim Shahir. In *Prehistoric Archaeology along the Zagros Flanks (vol. 105)*, ed. L. S. Braidwood, R. J. Braidwood, B. Howe, C. A. Reed and P. J. Watson. Chicago: The Oriental Institute of the University of Chicago, pp. 431-83.

Turnbull, P.F., 1983. The faunal remains from M'lefaat. In *Prehistoric Archaeology along the Zagros Flanks (vol. 105)*, ed. L. S. Braidwood, R. J. Braidwood,

B. Howe, C. A. Reed and P. J. Watson. Chicago: The Oriental Institute of the University of Chicago, pp. 693-5.

Turnbull, P. F.and Reed, C. A., 1974. The fauna from the Terminal Pleistocene of Palegawra Cave, a Zarzian occupation site in northeastern Iraq. *Fieldiana Anthropology* 63, 81-146.

Uerpmann, H. P., 1982. Faunal remains from Shams Ed-Din Tannira, a Halafian site in northern Syria. *Berytus* 30, 3-52.

Uerpmann, H.P., 1987. *The Ancient Distribution of Ungulate Mammals in the Middle East* (Beihefte zum Tübinger Atlas des Vorderen Orients: Reihe A (Naturwissenschaften) Nr.27). Wiesbaden: Dr.Ludwig Reichert Verlag.

Watkins, T. and Baird, D., 1987. *Qermez Dere. The excavation of an aceramic Neolithic settlement near Tel Afar, n. Iraq, 1987: interim report*. University of Edinburgh, Department of Archaeology Project Paper no. 6.

Watkins, T., Baird, D. and Betts, A., 1989. Qermez Dere and the early aceramic Neolithic of N. Iraq. *Paléorient* 15 (1), 19-24.

Watkins, T., Betts, A., Dobney, K., Nesbitt, M., Gale, R. and Molleson, T., 1991. *Qermez Dere, Tell Afar: Interim report no.2*. University of Edinburgh, Department of Archaeology Project Paper no 13.

Watkins, T., 1992. The beginning of the Neolithic: searching for meaning in material culture change. *Paléorient* 18, 63-75.

Watkins, T., Betts, A., Dobney, K. and Nesbitt, M. 1995. *Qermez Dere, Tel Afar: Interim Report No. 3*. University of Edinburgh, Department of Archaeology Project Paper no. 14.

THE FAUNAS OF THE PLEISTOCENE/HOLOCENE BOUNDARY IN THE SENO DE LA ULTIMA ESPERANZA, CHILE

Luis Alberto Borrero

Programa de Estudios Prehistóricos, Bartolomé Mitre 1970 - Piso 5 (1039), Buenos Aires, Argentina

Introduction

The goal of this paper is to discuss the chronology of ground sloth (*Mylodon darwinii*) at Cueva del Mylodon, Ultima Esperanza, Chile (51°36'S., 72° 36'W.). This is a huge cave of about 10,000 square meters, located some five kilometers from the Pacific Ocean. Excellent preservation of ground sloth skin, hair, muscular tissue and bone was recorded in the cave (Favier Dubois and Borrero 1997). The quality of preservation was attested by the recovery of DNA in fragments of ground sloth bone (Pääbo 1993: 88; Höss et al. 1996). This preservation was explained as the result of freezing (Sutcliffe 1985), but probably a combination of dryness and extreme cold, plus the sealing effect of salts must be taken into account (Favier Dubois and Borrero 1997: 212).

Systematic research at Cueva del Mylodon was initiated only a few years after its discovery in 1895. Erland Nordeskiold (1900, 1996) and Theodore Hauthal (1899) were already working in the cave by 1899. Unfortunately, due to the fantastic preservation of skin, the deposits of the site were also heavily searched for skin fragments to be sold to museums and individuals (Martinic 1996). For that reason, when the second wave of research started in the 1930s, conditions for research were very difficult, with the surface of the cave looking like a mining camp.

Several research teams studied the ground sloth remains during a period spanning more than 40 years. Dozens of radiocarbon dates on ground sloth remains were produced. Accordingly, laboratory techniques changed from the first sample - processed by Willard Libby himself - to the more recent accelerator dates. Comparability of the results, then, is difficult to achieve.

Three classes of remains were dated, dung, skin and bone. We will discuss the results separately. All the results are presented as uncalibrated radiocarbon dates. Excluded from this analysis are several Holocene dates, which apply strictly to the recent natural and cultural history of the deposits. None of the Holocene dates was made on specimens of ground sloth or any other Pleistocene fauna.

The Dung Dates

Near the end of the XIX Century at least one quarter of the cave floor was covered by ground sloth dung, with accumulations in excess of two meters. Thus, it is not surprising that dung was usually selected for dating. Nineteen ground sloth dung samples were dated (Table 7.1).

However, only seven have clear stratigraphic provenience. Recent work suggest that - in spite of a long history of site destruction - stratigraphic integrity is preserved in at least two loci within the cave, Trench 5 (Saxon 1979) and Profile 2 (Borrero et al. 1991: Figure 1).

Trench 5 was excavated by Saxon in 1976. It was located near the trench dug by the French Mission in 1953 (Emperaire and Laming 1954) and not too far away from the place where Junius Bird obtained the first sample to be dated from this cave (Bird 1988).

Table 7.1. Cueva del Mylodon. Dates on ground sloth dung.

Date	Laboratory	Provenience	Source
12496±148	BM-1209	Trench 5, Layer 1	Burleigh and Matthews 1982; Saxon 1979
12552±128	BM-1375	Trench 5, Layer 10	Saxon 1979
12308±288	BM-1210B	Trench 5, Layers 14-15	Burleigh and Matthews 1982; Saxon 1979
10880±300	GX-6243	Unstratified	Markgraf 1985
12020±460	GX-6244	Unstratified	Markgraf 1985
12285±480	GX-6245	Unstratified	Markgraf 1985
11775±480	GX-6246	Unstratified	Markgraf 1985
11905±335	GX-6247	Unstratified	Markgraf 1985
10575±400	GX-6248	Unstratified	Markgraf 1985
12270±350	A-2445	Unstratified	Markgraf 1985
13470±189	A-2446	Unstratified	Markgraf 1985
12240±150	A-2447	Unstratified	Markgraf 1985
12870±100	A-2448	Unstratified	Markgraf 1985
13560±180	A-1390	Unstratified	Long and Martin 1974
10812±325	LP-34	Hauthal Collection	Figini et al. ms
11330±140	LP-255	Profile 2, -16/-24 cm	Borrero et al. 1991
12570±160	LP-257	Profile 2, -82/-91 cm	Borrero et al. 1991
10200±400	SA-49	Trench in center of cave	Delibrias et al. 1964; Emperaire et al. 1961
10832±400	C-484	"upper manure layer"	Bird 1988; Bird ms; Arnold and Libby 1951

Three radiocarbon dates spanning some 1.5 meters cannot be statistically separated, suggesting that the base of the dung deposit is well preserved in that portion of the cave. The average is around 12.4 ka BP.

Profile 2 was excavated by Borrero and collaborators in 1989, as part of a project centered in the discussion of the Holocene survival of ground sloth, as a result of which previous interpretations about mid-Holocene presence of sloths were refuted. Two radiocarbon dates from the bottom and the top of the dung pile were respectively dated 12.5 and 11.3 ka BP (Borrero et al. 1991).

Bird described the location of recovery of his sample (C-484) as over "the edge of an old pit" (Bird 1988: 227) in "the upper portion of the manure layer" (Bird MS). This sample, then is in chronological order with the results from the two sequences. The chronological result of the sample recovered by the French Mission (SA-49) also suggests that the upper portion of the dung layer was reached.

Theodore Hauthal's sample (LP-34), obtained in 1899 and dated in the 1970s, also applies to the upper part of the dung layer. Effectively, in describing his work at the cave, Hauthal emphasized that he only completed shallow excavations (Hauthal 1899: 411).

The stratigraphic, systematically recovered dung samples span the period 11.3-12.5 ka BP. Three samples recovered separately by different projects - Hauthal, Bird and the French Mission - in three different settings of the cave, produced results around 10.2-10.8 ka BP. In addition, 11 unstratified samples collected by Paul S. Martin (see Long and Martin 1974; Markgraf 1985) produced results between 13.4 and 10.8 ka B.P. We do not know if they were recovered in stratigraphic order. However, together with the systematically recorded samples they suggest that the accumulation period for dung is extremely well constrained.

The Skin Dates
In spite of the fact that skin was the most spectacular find at the cave, only three dates were made on that material (Table 7.2). None of them was made on a specimen with a well recorded provenience within the deposits. The three dates span the period between 13.5 and 10.4 ka BP.

The Bone Dates
Only four dates were made on ground sloth bone from this cave (Table 7.3), and all of them fall between 13.2 and 10.3 ka B.P.A vertebra recovered by Erland Nordenskiold in his excavation of 1899 was recently analyzed (LU-749).

Nordenskiold's was the first systematic stratigraphic work on the cave, - and one of the first in the Americas as well.

Table 7.2. Cueva del Mylodon. Dates on ground sloth skin.

Date	Laboratory	Source
13500±410	R-4299*	Long and Martin 1974
10400±330	A-1391	Long and Martin 1974
13040±300	W-2998	Meyer Rubin, in Martinic 1996

*also reported as NZ-1680 (Saxon 1976)

He excavated the dung layer, where he found several ground sloth bones. The results of the radiocarbon analysis suggest that he reached the base of the dung layer.

On the other hand, the results of the bone sample recovered by Hauthal tend to confirm the fact that he only reached the upper part of the dung layer.

Discussion
In sum, the 26 dates obtained from dung, skin and bone from a single cave are comprised between 13.5 and 10.2 ka BP, clearly suggesting that sloths ceased to use the cave near 10.2 ka BP.

If we compare this results with those obtained in other sites in Ultima Esperanza the picture is the same (Table 7.4). In the first place, all the rest of the samples from Ultima Esperanza are bones, since none of the other sites of the region replicated the preservation of Cueva del Mylodon. Seven samples from four different caves fall between 13.4 and 11.3 ka BP.

Cueva del Mylodon, Cueva Lago Sofía 4 and Dos Herraduras 3 are all paleontological sites and provide information on the availability of sloths in Ultima Esperanza. Cueva del Medio and Cueva Lago Sofía 1 are archaeological sites, where hearths, artifacts and ground sloth, horse and camelid bones with cut marks were found (Nami 1987; Prieto 1991). Both sites are dated between 11.0 and 10.0 ka BP. They attest to the human use of sloths near the end of the Pleistocene during a period of overlap of around 1,000 radiocarbon years. At both sites, with dates of 12.7 and 12.9 ka BP respectively, there is evidence of the presence of sloths before the arrival of humans. The context of two dates of 11.9 and 11.5 ka BP from Cueva del Medio still is not published.

Table 7.3. Cueva del Mylodon. Dates on ground sloth bone collagen.

Date	Laboratory	Provenience	Source
12984±76	BM-728	British Museum Collection	Burleigh et al. 1977
13183±202	BM-1208	Trench 2, Layer 10	Saxon 1979
13260±115	LU-794	Layer C; vertebra;Nordenskiold Collection	Hákansson 1976
10377±481	LP-49	Fragment of skull; Hauthal Collection	Figini et al. ms

Table 7.4. Ultima Esperanza, other sites. Dates on ground sloth bone collagen.

Site	Date	Laboratory	Bone	Provenience	Source
Cueva Lago Sofía 1	12990±490	PITT-0939		c. 45 cm	Prieto 1991
Cueva Lago Sofía 4	11590±100	PITT-0940	Vertebra	-30 cm	Borrero et al 1997
Cueva Lago Sofía 4	11050±60	NSRL-3341	Dermal ossicle		T. Stafford, pers. comm.
Cueva Lago Sofía 4	13400±90	AA-11498	Dermal ossicle	-30 cm	Borrero et al 1997
Cueva del Medio	11990±100	AA-12577	Vertebra	-75/80 cm	Martinic 1996
Cueva del Medio	11570±100	AA-12578	Fragment	-75/80 cm	Martinic 1996
Cueva del Medio	12720±300	NUTA-2341	Fragment	Profile	Nami and Nakamura 1995
Dos Herraduras 3	11380±150	LP-421	Rib	-40 cm	Borrero, in prep.
Dos Herraduras 3	12825±110	AA-12574	Dermal ossicles	-35/-45 cm	Martinic 1996

The chronological evidence from the rest of Patagonia is very scanty. In Northern Patagonia there is a single dated paleontological location, which indicated that sloths were available between 12.6 and 10.8 ka BP (see Nami 1996). The oldest archaeological sites in that part of Patagonia are dated around 9.0 ka BP (Crivelli et al. 1993; Crivelli et al. 1996). In some of these sites sloth bones were found, but never in a context suggesting human exploitation. One way to look at this evidence is that humans arrived to the area inmediately after the extinction of sloths. A similar case can be defended for Baño Nuevo, Aysen, Chile, where a radiocarbon date of 11.4 ka BP was obtained for a sloth dermal ossicle. The same deposit contained archaeological evidence dated around 9.0 ka BP (Mena and Reyes 1998).

The cause for the disappearance of sloths is still debated (Borrero 1997). A deterioration in the quality of fodder was suggested as the main cause for their extinction (Salmi 1955), a claim not substantiated by subsequent analyses (Saxon 1976). Changes in diet forced by paleoenvironmental shifts were also considered (Markgraf 1985), but we still do not know the vegetal richness at the times near the extinction, as the analysis of pollen from new samples of ground sloth dung continues to expand the list of dominant plants (Moore 1978; Heusser et al. 1994). Coevolutionary disequilibrium has also been invoked (Graham and Lundelius 1984), but more dates on a variety of Pleistocene bones are needed in order to assess the importance of this process in South America.

What the chronological data for Cueva del Mylodon indicate is that the number of dates on sloth remains diminishes for the period following the explosive eruption of the Reclus volcano, located some 40 kilometers north of the site (Stern 1992), suggesting either a direct impact on the faunas - the remains of at least one sloth were recovered within a tephra layer at Dos Herraduras 3 - or in the vegetation. Anyway, if this was the case the sloth population recovered, as demonstrated by the number of dates on dung falling between 11.0 and 10.0 ka BP.

The classic alternative to the climatic and ecological hypotheses is human overkill (Martin 1973). Detailed chronological analyses, like those presented here for Cueva del Mylodon and other sites in Ultima Esperanza, will play an important role in the assessment of this hypothesis.

It was observed that there are at least a couple of indications of the extinction of sloths inmediately before the arrival of humans. It may well be that humans collaborated in the process of extinction only in some regions, specifically in those that were colonized before the end of the Pleistocene, but at many places - including most of Patagonia - other causes will need to be considered.

References

Arnold, J.R. and W.F. Libby, 1951. Radiocarbon Dates. *Science* 113, 111-120.

Bird, J., MS. Chile, 14C. Deposited at the American Museum of Natural History, New York.

Bird, J., 1988. *Archaeology and Travels in Central Chile.* Iowa City: Iowa University Press.

Borrero, L.A., 1997. The Extinction of the Megafauna: A Supra-Regional Approach. *Anthropozoologica* 25-26, 209-216

Borrero, L.A., J.L. Lanata and P. Cárdenas, 1991. Reestudiando cuevas: nuevas excavaciones en Ultima Esperanza. *Anales del Instituto de la Patagonia* 20, 101-110.

Borrero, L.A., F.M. Martin and A. Prieto, 1997. La Cueva Lago Sofía 4, Ultima Esperanza, Chile: Una madriguera de felino del Pleistoceno tardío. *Anales del Instituto de la Patagonia* 25, 103-122.

Burleigh, R. And K. Matthews, 1982. British Museum Natural Radiocarbon Measurements XIII. *Radiocarbon* 24, 151-170.

Burleigh, R., A. Hewson and N. Meeks, 1977. British Museum Natural Radiocarbon Measurements IX. *Radiocarbon* 19, 143-160.

Crivelli, E.A., D. Curzio and M. Silveira (1993). La estratigrafia de la cueva Traful 1 (Provincia del Neuquén). *Praehistoria* 1, 9-160.

Crivelli, E.A., U. Pardiñas, M.M. Fernández, M. Bogazzi, A. Chauvin, V, Fernández and M. Lezcano, 1996. Cueva Epull·n Grande (Pcia. Del Neuquén). Informe de avance. *Praehistoria* 2, 185-265.

Delibrias, G., M.T. Guillier and J. Labeyrie, 1964. Saclay Natural Radiocarbon Measurements I. *Radiocarbon* 6, 233-250.

Emperaire, J. and A. Laming, 1954. La grotte du Mylodon (Patagonie occidentale). *Journal de la Société des Américanistes* 43, 173-205.

Favier Dubois, C. and L.A. Borrero, 1997. Geoarchaeological Perspectives on Late Pleistocene Faunas from Ultima Esperanza Sound, Magallanes, Chile. *Anthropologie* 25(2), 207-213.

Figini, A.J., J.E. Carbonari, G.J. Gómez, R.A. Huarte and A.C. Zubiaga, MS. Dataciones radiocarbónicas de restos de la Cueva del Mylodon, Seno Ultima Esperanza.

Graham, R. and E. Lundelius, 1984. Coevolutionary Disequilibrium and Pleistocene Extinctions. In *Quaternary Extinctions*, ed. R. Klein and P.S. Martin. Tucson: The University of Arizona Press, pp. 223-249.

Hákansson, S., 1976. University Lund Radiocarbon Dates IX. *Radiocarbon* 18, 290-320.

Hauthal, T., 1899. Reseña de los hallazgos en las cavernas de Ultima Esperanza. *Revista del Museo de La Plata* 9, 409-420.

Heusser, C.J., L.A. Borrero and J.L. Lanata, 1994. Late Glacial Vegetation at Cueva del Mylodon. *Anales del Instituto de la Patagonia (Serie Ciencias Naturales)* 21, 97-102.

Höss, M., A. Dilling, A. Currant and S. Pääbo, 1996. Molecular Phylogeny of the Extinct Ground Sloth *Mylodon darwinii*. *Proceedings of the National Academy of Sciences* 93, 181-185.

Long, A. and P.S. Martin, 1974. Death of American Ground Sloths. *Science* 186, 638-640.

Markgraf, V., 1985. Late Pleistocene Faunal Extinctions in Southern Patagonia. *Science* 228, 1110-1112.

Martin, P.S., 1973. The Discovery of America. *Science* 179, 969-974.

Martinic. M., 1996. La Cueva del Milodon: historia de los hallazgos y otros sucesos. Relación de los estudios realizados a lo largo de un siglo (1895-1995). *Anales del Instituto de la Patagonia* 24, 43-80.

Mena, F. And O. Reyes, 1998 (in press). Montículos y cuevas funerarias en Patagonia: una vision desde Cueva Baño Nuevo-1, XI Region. *Actas del III Congreso Mundial de Estudios sobre Momias*, Universidad de Tarapacá, Arica.

Moore, D.M., 1978. Post-glacial vegetation in the South Patagonian territory of the giant ground sloth *Mylodon. Botanical Journal of the Linnean Society* 77, 177-202.

Nami, H.G., 1987. Cueva del Medio: perspectivas arqueológicas para la Patagonia austral. *Anales del Instituto de la Patagonia* 16, 103-109.

Nami, H.G. and T. Nakamura, 1995. Cronología radiocarbónica con AMS sobre muestras de hueso procedentes del sitio Cueva del Medio (Ultima Esperanza, Chile). *Anales del Instituto de la Patagonia* 23, 125-134.

Nami, H.G., 1996. New Assessments on Early Human Occupations in the Southern Cone. In *Prehistoric Mongoloid Dispersals*, ed. T. Akazawa and E.J. Szathmáry, Oxford: Oxford University Press, pp. 254-269.

Nordenskiold, E., 1996. Observaciones y descubrimientos en cuevas de Ultima Esperanza en Patagonia occidental. *Anales del Instituto de la Patagonia* 24, 99-123

Pääbo, S. (1993). Ancient DNA. *Scientific American*, November, 86-92.

Prieto, A., 1991. Cazadores tempranos y tardíos en la Cueva Lago Sofía 1. *Anales del Instituto de la Patagonia* 20, 75-100.

Salmi, M., 1955. Additional Information on the Findings in the Mylodon Cave at Ultima Esperanza. *Acta Geographica* 14, 314-333.

Saxon, E.C., 1976. La prehistoria de Fuego-Patagonia: colonización de un habitat marginal. *Anales del Instituto de la Patagonia* 7, 63-73.

Saxon, E.C., 1979. Natural prehistory: the archaeology of Fuego-Patagonian ecology. *Quaternaria* 21, 329-356.

Stern, C., 1992. Tefrocronología de Magallanes: nuevos datos e implicaciones. *Anales del Instituto de la Patagonia* 21, 129-141.

Sutcliffe, A.J. (1985). *On the Track of Ice Age Mammals.* Cambridge: Harvard University Press.

ARCHAEOLOGICAL AND ZOOARCHAEOLOGICAL REMAINS FROM THE AUCILLA RIVER, NORTHWEST FLORIDA

Tanya M. Peres
Department of Anthropology, University of Florida, Gainesville, FL 32601

Brinnen S. Carter
Southeastern Archaeological Center, National Park Service, Tallahassee, FL 32303

Sites along the Aucilla River in Northwest Florida have provided ample evidence of the association of humans with both Late Pleistocene and Early Holocene fauna. This paper introduces one of those sites – the Page/Ladson site – by discussing its chronology, stratigraphy, zonation, associations, radiocarbon dates, and faunal assemblage.

Environmental Context of the Page/Ladson Site

Page/Ladson lies in a belt of karst sinks and rises that form a part of the Aucilla River (Figure 8.1). In this belt, the Aucilla flows underground for segments as long as five kilometers. It also flows in above-ground sections that range in size from 30 meter to over two kilometers. Page/Ladson is located in the two kilometer-long "Half Mile Rise" section of the river. The river is surrounded by large expanses of flat terrain; the maximum difference in elevation from sea level to the highest point is three meters.

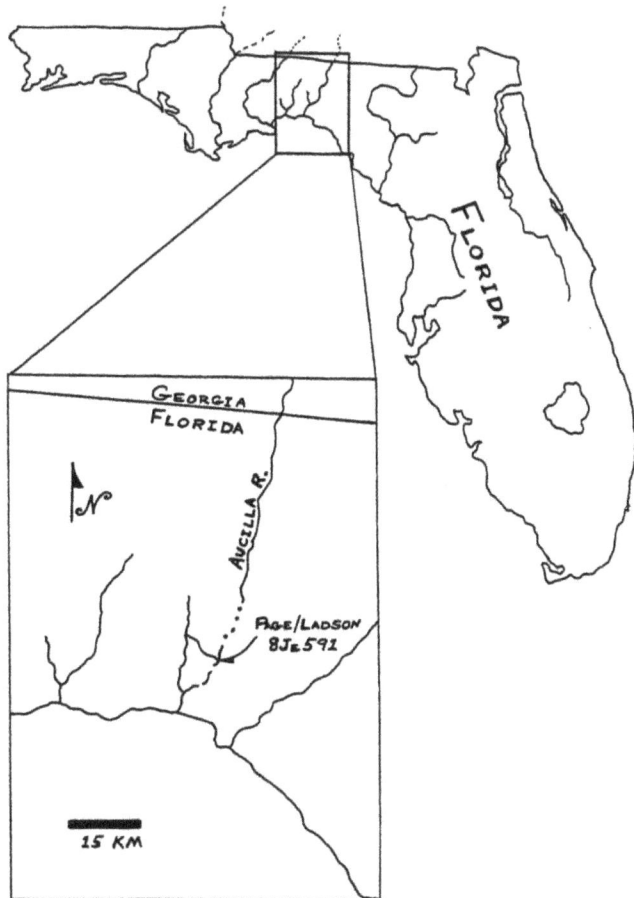

Figure 8.1. Location of Aucilla River and the Page/Ladson site.

Soils of the surrounding area are typically sands to sandy clays. Most are covered with a 10cm to 20cm layer of actively decomposing organic matter. These top soils are normally acidic, ranging from approximately pH 5 to 7. The Page/Ladson site itself is located at the confluence of the Aucilla, which is fed by runoff from adjacent upland swamps, and the spring-fed Wacissa River. The flow of the two combined is about 500 cubic feet per minute (CFM). Seasonal leaf-drops, and flooding, deposit and redeposit a leaf molt that ranges from 0m to 4m in depth on the bottom of the river. The twiggy leaf molt layers are typically interspersed with sand lenses. Thicker deposits are found in areas of low stream flow – mostly in deeper "sink" areas. Thinner deposits are found over bedrock limestone and dolomite shoals. Page/Ladson is on the lee side of one of the sink areas – in an eddy – and is not subject to annual erosion. To a certain extent, the annual leaf-molt deposits have protected the Late Pleistocene and Early Holocene deposits from erosion during the Late Holocene.

Modern ecosystems of the area include coastal tide marshes, cypress-dominated river swamps and bayheads, hardwood-dominated river swamps, pine-dominated flatwoods and slash pine plantations, upland hardwood hammocks, and over-drained pine-dominated sand hills. Local fauna include most species that typically inhabit these ecosystems, including relatively rare mammals such as black bear (*Ursus americanus*), Florida panther (*Felis concolor*), and beaver (*Castor canadensis*).

Paleoenvironment of the Lower Aucilla River

For the purposes of this paper, the Pleistocene/Holocene boundary is held to be at 10,000 BP. Several recent papers (Grimm and Jacobsen 1992; Watts et al. 1996) recognize that there is a peak in the rate of pollen change almost precisely at 10,000 BP, coinciding with changes in the stable oxygen isotope ratios in the GRIP and GISP2 cores and with changes in available atmospheric Carbon-14. At 10,000 BP acceleration of sea level rise occurs that does not cease until around 7,000 BP (Milliman and Emery 1968).

During the period covered in this paper, the sea was approximately 125km from the Page/Ladson site. Local ecosystems were substantially different. Pollen from the 10,000 year old levels indicate that the surrounding uplands were oak-dominated, with abundant grasses and riverine stands of cypress (Watts 1971). Contemporaneous fauna will be discussed later in this paper.

Because the period around 10,000 BP is one of rapid, dramatic environmental change, it seems appropriate to refine existing archaeological chronologies so that the effect of this change on human and animal populations can be evaluated. Evaluating archaeological sites from both sides of the transition together effectively defeats any attempt to confirm or deny that the environmental change had an effect on any aspect of Early Archaic life. Similarly, evaluating fauna from both sides of the transition together prevents zooarchaeologists from accurately assessing faunal replacement or range shifts that might occur as a result of the general environmental restructuring.

Paleoindian and Early Archaic Periods in Florida

Although the definition of the difference between Paleoindian and Archaic has rested on several different criteria over the years, Carter has adopted a conservative interpretation that links Paleoindians to megafauna hunting. In Florida, the Paleoindian/Early Archaic transition appears to occur prior to the development of side-notched points, as side-notched points have not been found in good archaeological context with any Pleistocene mega-herbivore.

Thus, we are left with two options: 1) "What is the chronological position of side-notched or Early Archaic assemblages and what does Page/Ladson tell us about the assemblages' temporal placement?" and 2) "What does the fauna from Page/Ladson say about the available edible animals in the local area and their own deposition?" The answers lie in the Page/Ladson site materials themselves. We now review the Page/Ladson Late Pleistocene strata, dates, and selected diagnostic artifacts.

Page/Ladson Stratigraphy

The Page/Ladson site has both underwater-components and land-components. At the time the Early Archaic remains were deposited, the entire site was approximately 60m above mean sea level and was likely a streamside environment. The now-inundated deposits begin with the previously described leaf-molt and sand lenses. Immediately underlying these recent deposits is a reddish peat that yields ceramics dating to the Deptford period (500 BC - AD 100) (Milanich and Fairbanks 1980: 66). Below this are horizontally-bedded gray, organic clays that are truncated on the top and bounded on the bottom by a shelly hash. The clay/shell hash in turn overlies an organic-rich, degraded peat deposit. This deposit is colloquially called the "Bolen level" for the type of side and corner-notched points that have been found in and on this 10cm thick level. The diagnostic Bolen points (Bullen 1958, 1975) have been found in sealed associations with preforms, adzes, bolo stone preform fragments, and antler flakers and handles (Carter 1996). Immediately underlying the "Bolen level" are 2 to 3 meter-thick beds of organic-rich gray clays that have bounding dates of 10,000 BP and 11,700 BP. Immediately below these lower clays are sandy marls that have yielded extinct horse (*Equus* sp.), camel (*Camelidae*), and mammoth/mastodons (*Proboscidea*).

Focusing on the "Bolen level," four radiocarbon dates on wood stakes, a large log, and a hickory nut, all cluster around 10,000 BP. Sediment composition, artifact deposits, and several "hearths," suggest that the site was occupied for a very limited time immediately prior to 10,000 BP, when the site became inundated. The only diagnostic artifacts from these levels are notched lithics.

The dates from Page/Ladson suggest that side-notched points were developed before 10,100 BP, potentially during a period

of relative environmental stability. Also, if Goodyear (1982) is correct in positing a 9,900 BP date for the end of the manufacturing of Dalton points in the Southeast, then there is limited evidence that Florida groups abandoned lanceolate points before groups in the rest of the Southeast. Recent evidence from Dust Cave, in northern Alabama (Driskell 1995), appears to largely confirm this scenario. Early in the excavations at Dust Cave, Driskell noted the presence of side-notched points in a stratum that had previously only yielded lanceolate points. Driskell labeled them intrusive by means of water movement in the back of the cave. Could they have been part of the lanceolate tool kit instead?

Early Archaic Fauna at Page/Ladson

Fauna from the Early Archaic levels at Page/Ladson provide a measure of both the available food resources in the area and whether particular deposits were laid down by natural or cultural forces.

In analyzing the Page/Ladson assemblage, one of the main concerns that was addressed was the depositional nature of the remains. Were these remains there as a result of human activities or environmental conditions? The zooarchaeological literature was turned to for guidance in addressing this issue. After reviewing the literature, a set of observational criteria was chosen, and a scoring system was developed as an aid in determining the nature of the faunal assemblage. The criteria that were used to determine if an assemblage has been manipulated by humans are: the presence of butchering marks or evidence of working/use-wear; the assemblage's location within the stratigraphic sequence; and the assemblage's relationship to features and artifacts. Taphonomic history, excavation context and associations, stratigraphic integrity, and investigator bias are factors that were also considered, but not included in the actual "scoring" of an assemblage.

In studying faunal remains that may be associated with archaeological or paleontological contexts, it is important to know what types of biases may affect the sample. Two types of bias are most common: 1) those introduced as a result of the taphonomic history, and 2) those resulting from the methods of archaeological recovery.

The taphonomic history of an assemblage determines the extent of the preservation of those animals in the fossil or archaeological record. Within zooarchaeology and paleontology, taphonomic studies are used to understand the agents that have resulted in the particular assemblage's preservation, and to gain a perception of what has been lost. Differential preservation, one of the many taphonomic processes, plays a significant role in determining what will remain among assemblages that are approximately 10,000 years old. Many late Pleistocene sites, especially those located on land, do not have a representative faunal assemblage because of poor preservation conditions of the environment, e.g. Harney Flats in Florida (Daniel and Wisenbaker 1987). This is not so at the Page/Ladson site, where remains are excellently preserved and recovered in great numbers. Investigator bias must be taken into consideration because the analyst must know to what degree, beyond mere taxonomic identification, the faunal sample may be confidently analyzed.

In evaluating the type of faunal assemblage present, one must consider the type of site from which the materials were recovered, association with features and artifacts, origin of faunal species, skeletal completeness, presence/absence of butchering marks, worked bone, and burnt bone. In preparing a model of an expected Page/Ladson faunal assemblage, all of the above criteria were considered. These criteria, and information on the site's environment, enabled the status of the assemblage to be evaluated by employing a scoring system (Table 8.1). This scoring system allows the analyst to place an assemblage in a group by assigning points for various characteristics that may be present (or absent) in the sample. This system is not meant to replace a consideration of all assemblage attributes, but it can aid in focusing the research. Skeletal completeness is not included in the system, since it is a criterion that can have different interpretations, particularly where context is concerned. After an assemblage has been scored, the cultural component of the site, the site type (shell midden, camp, etc.) and the degree of stratigraphic control must be considered.

Table 8.1. Scoring system for faunal assemblages (Peres 1997)

Criteria	Points	
	Yes	No
Presence of butchering marks	2	0
Presence of worked bone	2	0
Association with features/artifacts	2	0
Evidence of thermal alteration	1	0
Presence of "exotic" taxa	1	0
Environmental deposit	0 - 2 points	
Indeterminate deposit	3 - 5 points	
Cultural deposit	6 - 8 points	

As can be seen from Table 8.1, the first three criteria have higher point values, as they are the most effective measures in determining the assemblage to be cultural. Thermal alteration and presence of "exotic" taxa do not always imply human manipulation; thus they each have lower point values. In many cases, the scores reflect the depositional nature of the site. In other cases, the scores reflect an indeterminate depositional nature, thus requiring one to turn to other avenues of information, such as the type of site, degree of stratigraphic control, and for the remains themselves, skeletal completeness.

The underwater excavations at Page/Ladson were initiated by divers removing the overburden of leaves, sand, clay, and peat, aided by a six-inch water dredge from six one m^2 units labeled O, P, Q, T, U, V. These materials were screened through _ inch mesh, and given the label of "General Surface Collection." The overburden was removed to approximately 20 cm above the "Bolen level" (ARPP 1995).

From this point, the divers excavated each unit in 10 cm, arbitrary levels within the natural stratigraphy. This material was removed by hand-fanning, aided by a four-inch water dredge. The excavated material was recovered using _ inch hardware cloth. When the excavators reached the zone immediately preceding the Bolen surface (Zone 3), material was screened through 1/8-inch mesh. Material was removed until the top of the Bolen level (Zone 4) was exposed. Divers piece plotted artifact such as wood, lithics, and fauna that were located on this surface, then removed them.

Each zone was recognized by characteristics of the soil matrix. The excavation logs describe the soil of Zone 3 (immediately on top of the Bolen surface) as being a soft grey clay containing an abundance of small snail shells, a fair number of organic materials (small twigs, branches, root casings), and minute amounts of limestone rock and charcoal fragments.

The first visual sort of the remains entailed dividing them into classes within each field specimen lot. Next elements were identified to the appropriate taxa using the Zooarchaeological Comparative Collection of both the Florida State University and the Florida Museum of Natural History. In this collection the assemblage can be arbitrarily divided into several smaller vertebrate faunal groupings based on method of recovery. Data will be presented in the following manner: Group 1: the four arbitrary 10cm levels within Zone 3 for all Units (O, P, Q, T, U, V); Group 2: individual general surface collections; Group 3: piece plotted specimens; and Group 4: modified bone; three pieces of modified bone were handpicked from the surface, but not piece-plotted. Each of the four groupings was scored to determine the nature of their deposition.

After scoring Group 1 (Table 8.2), it appears that the material recovered from all six units (Zone 3, Levels 1-4) is the result of environmental deposition. This conclusion is reinforced by other factors, the most significant of which is that the faunal elements do not show evidence of modifications, and many can be articulated. Relative size is also important and many of the taxa present are small (e.g., Kinosternidae). The scores of Group 2 (Table 8.2) indicate that the assemblage was indeed a result of environmental deposition. Such a finding seems reasonable given that the surface collections were taken form

Table 8.2. Groups 1 - 4 assemblage characteristics

Sample	Butchering Marks	Modified Bone	Assoc. Features/ Artifacts	Thermal Alteration	Exotic taxa	Total Points	Deposit Type
Group 1: Zone 3, Levels 1-4	N	N	N	N	N	0	Environmental
Group 2: Surface Collections	N	N	N	N	N	0	Environmental
Group 3: Piece Plotted Items	N	Y	Y	N	N	4	Indeterminate
Group 4: Modified Bone	Y	Y	Y	N	N	6	Cultural

the same zone and levels as Group 1. The scoring of Group 3 indicates that the depositional circumstances are indeterminate. There is an artifact in this assemblage (a possible bone pin) that is clearly the result of human manipulation. The results of scoring Group 4 indicate that these materials were deposited in the Aucilla area as a result of human presence. Two bone pins and a half rack of antler with cut marks are indisputable examples of the kinds of modifications expected in a cultural assemblage.

It is apparent from the scores of the four groups that the majority of this collection of vertebrate faunal material from the Page/Ladson site was deposited as a result of environmental agents. These animals lived in and around this prehistoric watering hole, and many of them died there, not always as a result of human actions. Only one group (Group 4) in the Page/Ladson sample had a cultural score because the three specimens in this group were clearly modified by humans. This group is positively cultural in *origin* but possibly not in deposition. The presence of these artifacts may be a result of redeposition during the early Holocene. They may have been transported downstream or washed in from the land.

In conclusion, it appears that important transformations in the local environments occurred around 10,000 BP.

Sediments and fauna appear to have washed into the inundated portion of the Page/Ladson site, probably as a result of early Holocene flooding. However, the worked bone from in and on the "Bolen level" likely indicates that humans were either living on the surface on or before 10,000 BP, or were discarding broken artifacts onto the surface. Further analysis should resolve this question. Future studies at Page/Ladson should include intentional sampling (such as column, flotation, and smaller mesh sizes) for full recovery of fauna and flora remains. Additional studies to aid in this effort should use the same types of recovery strategies from other sites located along the Aucilla River, and along other waterways in Florida. Faunal collections from previous field season, and those recovered since the 1995 excavations should be analyzed with similar questions in mind and added to this database.

Acknowledgements
The authors would like to thank the Aucilla River Prehistory Project, the Florida Department of State's Bureau of Historic Preservation Special Category Grants Program, and the University of Florida Department of Anthropology for supporting this study.

References

Aucilla River Prehistory Project (ARPP), 1995. Original unpublished field notes from the Page/Ladson site. Ms. on file, Gainesville: Florida Museum of Natural History.

Bullen, R. P., 1958. *The Bolen Bluff site on Paynes Prairie, Florida*. Gainesville: Contributions to the Florida State Museum, no. 4.

Bullen, R. P., 1975. *A Guide to the Identification of Florida Projectile Points*. Gainesville: Kendall Books.

Carter, B. S., 1996. The Bolen surface: a story of opal and precious stones. *The Aucilla River Times* 9(1), 8-9.

Driskell, B. N., 1995. Stratigraphy and chronology at Dust Cave. *Journal of Alabama Archaeology* 40(1), 17-34.

Daniel, I. R., Jr. and M. Wisenbaker, 1987. *Harney Flats: A Florida Paleo-Indian Site*. Farmingdale: Baywood Publishing.

Milanich, J. T. and C. H. Fairbanks, 1980. *Florida Archaeology*. New York: Academic Press.

Goodyear, A., 1982. The chronological position of the Dalton horizon in the Southeastern United States. *American Antiquity* 47(2), 382-395.

Grimm, E. C. and G. L. Jacobson, Jr., 1992. Fossil-pollen evidence for abrupt climatic changes during the past 18,000 years in eastern North America. *Climate Dynamics* 6, 179-184.

Milliman, J. D. and L. O. Emery, 1968. Sea-levels during the past 35,000 years. *Science* 162, 1121-1123.

Peres, T. M., 1997. Analysis of a late Pleistocene and early Holocene faunal assemblage from the Page/Ladson site (8JE591), Jefferson County, Florida. Unpublished Masters Thesis. Department of Anthropology, Florida State University.

Watts, W. A., 1971. Postglacial and interglacial vegetation history of Southern Georgia and Central Florida. *Ecology* 52, 676-690.

Watts, W. A., E. C. Grimm, and T. C. Hussey, 1996. Mid-Holocene forest history of Florida and the Coastal Plain of Georgia and South Carolina. In *Archaeology of the Mid-Holocene Southeast*, ed. K. E. Sassaman and D. G. Anderson. Gainesville: University of Florida Press, pp. 28-38.

SHERIDEN: A STRATIFIED PLEISTOCENE-HOLOCENE CAVE SITE IN THE GREAT LAKES REGION OF NORTH AMERICA

Kenneth B. Tankersley

Department of Anthropology, Kent State University, Kent, Ohio, 44242, USA

Sheriden Cave

Sheriden is a transitional late Pleistocene-early Holocene cave site located in the Lake Erie basin of northwestern Wyandot County, Ohio, USA. The cave entrance is a funnel-shaped sinkhole situated atop a 17-m high bedrock ridge at an elevation of approximately 272 m above mean sea level (amsl). The ridge is partially buried by glacial deposits. If the these sediments were removed, the relief of the ridge would be more than to 30 m above the surrounding landscape (House 1985:78).

Sheriden Cave formed in a porous, stromatolitic, Upper Silurian, Salina Group, Greenfield Dolomite. Its surface has a granular appearance and saccharoidal texture. Massive, cabbage-head shaped stromatolites are exposed along the walls and ceilings of the cave. These structures contain abundant voids that promote movement of solvent ground waters. Like other caves in Ohio (Carlson 1991:65; Hoy et al. 1995:83), Sheriden formed as ground water widened voids in the dolomite through dissolution. Active water circulation hastened collapse, the formation of the main cave passage, and its opening to the surface (Tankersley 1997).

Although the exact time of cave formation is unknown, the depositional history of the cave is undoubtedly related to the evolution of the late Quaternary landscape. The last glacial advance over the site occurred when the Erie and Huron lobes of the Laurentide ice sheet coalesced during the Late Wisconsin, Late Woodfordian, Carey Stade, ca. 15,000 radiocarbon yr B.P. (Totten 1985). Sometime between 14,400 and 13,800 radiocarbon yr B.P., the edge of the Laurentide ice thinned and retreated out of Wyandot County and into the Lake Erie basin (Calkin and Feenstra 1985). Ice recession created four glacial lakes in Wyandot County: Lake Killdeer (> 280 m amsl), Lake Wharton (> 267 m amsl),

Lake Vanlue (> 253 amsl), and Lake Carey (> 252 m amsl). Temporally, they correlate with four glacial lakes in the Lake Erie Basin (i.e., Lakes Maumee I, II, III, and Lake Arkona respectively). The limited extent of lacustrine deposits, the lack of shoreline features, and the shallowness of their channels suggest that the glacial lakes were short-lived in the vicinity of Sheriden Cave (House 1985:75).

While subsequent glacial events did not directly impact the cave, climatic changes associated with the Port Huron (ca. 13,000 to 12,000 radiocarbon yr B.P.), and Great Lakean (ca. 11,800 to 10,800 radiocarbon yr B.P.) stades and Two Creekan interstade (ca. 12,000 to 11,800 radiocarbon yr B.P.) would have had profound effects on sedimentation within the cave. During this time, the region experienced dramatic fluxuations in the water table. Between ca. 11,400 and 10,400 radiocarbon yr B.P., groundwater in the Lake Erie basin plummeted to its lowest levels (Anderson and Lewis 1985:245; Barnett 1985:191). After ca. 10,000 radiocarbon yr B.P., the effects of glacial melting, climatic change, and isostasy caused the regional water levels to rise dramatically and eventually inundate the cave passages.

Unconsolidated Stratigraphy

Sheriden Cave contains a deep stratigraphic sequence of unconsolidated late Pleistocene and early Holocene deposits. Four major stratigraphic units (I-IV) have been defined on the basis of color and particle size analysis (Table 9.1). Unit I can be subdivided into two strata: an uppermost layer (Ia) of dark brown silt and fine sand-sized ceiling rain deposited after the horizontal cave sealed; and a lower stratum (Ib) of dark brown, low energy in-wash from the surface of the pit and horizontal cave mouth. Unit II includes three strata displaying some soil development: IIa, a dark brown, high-

Table 9.1. Particle size analysis of sediments from Sheriden Cave based on total dry weight.

Unit	Color (moist)	Percent Particle Size (mm)				
		Cobble Gravel >19.1	Granule Pebble 19.1-2.0	Sand 2.0-0.063	Silt 0.063-0.004	Clay <0.004
Ia	10YR 4/3	0.0	0.0	58.8	22.3	18.9
Ib	10YR 4/3	0.0	0.1	12.5	51.4	36.0
IIa	10YR 3/3	23.1	16.5	10.3	27.5	22.6
IIb	10YR 3/2	0.0	0.1	19.8	46.9	33.2
IIc	10YR 3/3	32.6	18.8	16.4	18.8	13.4
IIIa	10YR 5/3	47.5	28.7	13.7	6.2	3.9
IIIb	10YR 5/4	0.0	0.9	24.7	43.6	30.8
IIIc	10YR 5/4	25.9	16.7	26.6	19.2	11.6
IVa	7.5YR 6/8	0.0	4.4	28.0	11.1	56.5
IVb	10YR 6/4	0.0	0.2	0.6	13.3	85.9
IVc	10YR 6/4	0.0	0.0	13.6	72.8	13.6
IVd	10YR 6/4	0.0	0.0	8.0	52.3	39.7
IVe	10YR 4/2	0.0	0.0	1.4	31.2	67.3

energy clastic flow, presumably from the pit and cave mouth; IIb, a dark greyish brown low energy in-wash very similar to stratum Ib; and IIc, a dark brown, high-energy clastic flow that is virtually identical to stratum IIa. Unit III consists of three strata: IIIa, a brown high energy clastic flow of minimally altered till; IIIb, a yellowish brown, low energy in-wash similar to strata Ib and IIb except for color; and IIIc, a yellowish brown, medium-energy (or mixture of high and low energy) in-wash with a high clay-silt-fine sand ratio. Unit IV contains at least five strata. Unit IVa, a highly oxidized reddish yellow film, is a mixture of the upper-most lake clays and silt- and sand-sized ceiling rain. It represents a long period, perhaps more than 1,000 years, of no deposition. Units IVb, IVc, IVd, and IVe are light yellowish brown to dark greyish brown lake clays with higher silt and fine sands representing some periods of turbidity.

The unconsolidated stratigraphic record undoubtedly resulted from a complex and dynamic system of sedimentary processes that are directly related to the surface landscape. Under normal periods of deposition, that is, normal rainfall and snow melt runoff into the pit, there would have been a downslope fining and thinning of sediments. Such a depositional sequence is represented by Units Ib, IIb, and IIIb. However, during extremely heavy rains and snow melt runoff (e.g., ≥ 25 cm/24 hrs), sediment plugging the pit would have been blown-out with a simultaneous catastrophic collapse and flowage of the rim of sediments that momentarily resulted when the bottom of the pit went out. Units IIa, IIc, IIIa, and IIIc are indicative of such catastrophic clastic flows. In the clastic flow event represented by Unit IIIa, sediments from the pit subsided to well below the level of the horizontal cave, and the top of the sediments in the pit did not again reach the level of the horizontal cave for many years.

Vertebrate Paleontology
The disarticulated remains of more than 60 vertebrate taxa have been recovered from Units I, II, and III (Table 9.2) (Ford et al. 1996; Holman 1997; McDonald 1994; Tankersley 1997). With the exception of the masked shrew (*Sorex cinereus*), all of the vertebrates recovered from Unit I consist of amphibians, reptiles, and mammals that are living in the immediate vicinity of the cave today.

The bulk of the vertebrate taxa were recovered from Unit II including five extinct and eleven extralimital species. The extralimital masked shrew (*Sorex cinereus*), yellow-cheeked vole (*Microtus xanthognathus*), heather vole (*Phenacomys intermedius*), porcupine (*Erethizon dorsatum*), and extinct flat-headed peccary (*Platygonus compressus*) are ubiquitous in Units IIa, IIb, and IIc. At least 36 individuals of flat-headed peccary have been recovered Unit II. They are of varing age, with young adults of approximately 1 and 3-5 years predominating (Gobetz 1998:iv). Although the peccary skeletons appear to be incomplete, the small portion of bones with heavy abrasion, flaking, or rodent gnawing demonstrate that individuals entered the cave as whole carcasses and were buried in a state of minimal decomposition (Gobetz 1998:v). This position is further supported by the fact that some of the bones display tooth marks of large and small carnivores.

A single, heavily-weathered, and rodent gnawed antler of the extralimital caribou (*Rangifer tarandus*) was excavated from Unit IIa. Dentition and the osseous remains of at least one extinct giant beaver (*Castoroides ohioensis*) were excavated from Units IIa and IIb. Dentition and osseous remains of the northern bog lemming (*Synaptomys borealis*) were recovered from Units IIb, IIc, and IIIc. The extralimital pygmy shrew (*Sorex hoyi*), ermine (*Mustela erminea*), pine marten (*Martes americana*), fisher (*Martes pennanti*) and the dentition and osseous remains of a single extinct stag-moose (*Cervalces scotti*) and short-faced bear (*Arctodus simus*) were retrieved from Unit IIc. A third molar and right-side mandible of the long-nosed peccary (*Mylohyus naustus*) were recovered from disturbed contexts (Gobetz 1998:23).

Most of the extralimital mammals from Unit II are well south of their present ranges in boreal and tundra environments. They occur in direct stratigraphic association with mammals that have temperate or mid-latitude biogeographic distributions. The stratigraphic association of these presently allopatric species is an example of a non-analog association that seems to be characteristic of late Pleistocene vertebrate communities (Semken et al. 1998). In terms of paleoenvironmental reconstruction, the faunal assemblage is suggestive of a late Pleistocene mosaic habitat consisting of a shallow, marshy pond grading into an open woodland with a grassy ecotonal area (Holman 1997).

Table 9.2. Stratigraphic context of vertebrate taxa from Sheriden Cave.

Vertebrate taxa	Common name
Ia	
Microtus pennsylvanicus	Meadow Vole
Peromyscus maniculatus	Field Mouse
Marmota monax	Woodchuck
Ib	
Rana clamitans	Green Frog
Rana pipiens	Northern Leopard Frog
Nerodia sipedon	Northern Water Snake
Thamnophis sp.	Garter Snake
Cryptotis parva	Least Shrew
Sorex cinereus	Masked Shrew[a]
Myotis sp.	Bat
IIa	
Nocomis biguttatus	Hornyhead Chub
Notemigonus crysoleucas	Golden Shiner
Ambystoma laterale	Blue-spotted Salamander
Ambystoma jeffersonianum	Jefferson's Salamander
Bufo americanus	American Toad
Pseudacris triseriata	Striped Chorus Frog
Rana clamitans	Green Frog
Rana pipiens	Northern Leopard Frog
Rana sylvatica	Wood Frog
Chelydra serpentina	Snapping Turtle
Emydoidea blandingii	Blandings Turtle
Thamnophis sp.	Garter Snake
Opheodrys vernalis	Smooth Green Snake
Blarina brevicauda	Short-tailed Shrew
Sorex cinereus	Masked Shrew[a]
Leporidae sp.	Rabbit/Hare
Microtus pennsylvanicus	Meadow Vole

Microtus xanthognathus	Yellow-cheeked Vole[a]
Ondatra zibethica	Muskrat
Phenacomys intermedius	Heather Vole[a]
Peromyscus maniculatus	Field Mouse
Tamias striatus	Eastern Chipmunk
Tamiasciurus hudsonicus	Red Squirrel
Marmota monax	Woodchuck
Castor canadensis	Beaver
Castoroides ohioensis	Giant Beaver[b]
Erethizon dorsatum	Porcupine[a]
Procyon lotor	Raccoon
Ursus americanus	Black Bear
Platygonus compressus	Flat-headed Peccary[b]
Rangifer tarandus	Caribou[a]

IIb

Moxostoma macrolepidotum	Shorthead Redhorse
Ictalurus ponctatus	Channel Catfish
Ictalurus furcatus	Brown Bullhead
Campostoma anomalum	Central Stoneroller
Semotilus atromaculatus	Creek Chub
Micropterus dolomieu	Smallmouth Bass
Ambystoma laterale	Blue-spotted Salamander
Bufo americanus	American Toad
Bufo woodhousii	Fowler's Toad
Rana catesbeiana	Bull Frog
Rana pipiens	Northern Leopard Frog
Rana sylvatica	Wood Frog
Testudinae sp.	Turtle
Serpentes sp.	Snake
Elaphe vulpina	Fox Snake
Regina septemvittata	Queen Snake
Thamnophis sirtalis	Common Garter Snake
Myotis septentrionalis	Northern Bat
Leporidae sp.	Rabbit/Hare
Microtus pennsylvanicus	Meadow Vole
Microtus xanthognathus	Yellow-cheeked Vole[a]
Ondatra zibethica	Muskrat
Phenacomys intermedius	Heather Vole[a]
Synaptomys borealis	Northern Bog Lemming[a]
Peromyscus maniculatus	Field Mouse
Marmota monax	Woodchuck
Castor canadensis	Beaver
Castoroides ohioensis	Giant Beaver[b]
Erethizon dorsatum	Porcupine[a]
Procyon lotor	Raccoon
Platygonus compressus	Flat-headed Peccary[b]

IIc

Ambystoma laterale	Blue-spotted Salamander
Rana sylvatica	Wood Frog
Testudinae sp.	Turtle
Serpentes sp.	Snake
Sauria sp.	Lizard
Blarina brevicauda	Short-tailed Shrew
Sorex hoyi	Pygmy Shrew[a]
Sorex cinereus	Masked Shrew[a]
Myotis septentrionalis	Northern Bat
Leporidae sp.	Rabbit/Hare
Microtus pennsylvanicus	Meadow Vole
Microtus xanthognathus	Yellow-cheek Vole[a]
Ondatra zibethica	Muskrat

Phenacomys intermedius	Heather Vole[a]
Synaptomys borealis	Northern Bog Lemming[a]
Peromyscus maniculatus	Field Mouse
Tamias striatus	Eastern Chipmunk
Sciurus carolinensis	Eastern Gray Squirrel
Tamiasciurus hudsonicus	Red Squirrel
Marmota monax	Woodchuck
Glaucomys sabrinus	Northern Flying Squirrel
Castor canadensis	Beaver
Erethizon dorsatum	Porcupine[a]
Vulpes vulpes	Red Fox
Canis lupus	Gray Wolf
Procyon lotor	Raccoon
Arctodus simus	Short-faced Bear[b]
Martes americana	Pine Marten[a]
Martes pennanti	Fisher[a]
Mustela vison	Mink
Mustela ermina	Ermine[a]
Platygonus compressus	Flat-headed Peccary[b]
Cervalces scotti	Stag Moose[b]
Odocoileus virginianus	White-tailed Deer

IIIa
No Vertebrates
IIIb
No Vertebrates
IIIc

Synaptomys borealis	Northern Bog Lemming[a]

IVa
No Vertebrates
IVb
No Vertebrates
IVc
No Vertebrates
IVd
No Vertebrates
IVe
No Vertebrates
a. Extralimital Species
b. Extinct Species

Paleobotany

Coniferous and deciduous wood charcoal is abundant in the upper 30 cm of the late Pleistocene deposits. Some of the conifer wood charcoal has resin ducts and is indicative of quintessential late Pleistocene Great Lake species such as pine (*Pinus* sp.), spruce (*Picea* sp.), and larch (*Larix laricina*). However, most of the conifer charcoal lacks resin ducts, a characteristic of atypical late Pleistocene Great Lake species such as eastern red cedar (*Juniperus virginiana*), white cedar (*Thuja occidentalis*), hemlock (*Tsuga canadensis*), and balsam fir (*Abies balsamea*). The deciduous wood charcoal belongs to the Salicaceae family and includes willow (*Salix* sp.) and poplar (*Populus* sp.). Willows and poplars tend to grow in mesic environments suggesting that the area in and around the cave was moist. The co-occurrence of fish scales and Salicaceae charcoal demonstrates that water was present in the immediate vicinity of the sink entrance.

Archaeology

A number of distinctive Paleoindian artifacts were recovered from Unit II (Table 9.3). A split-bone projectile point and a cervical vertebra of a snapping turtle (*Chelydra sepentina*) displaying distinctive chopping and cutmarks were exposed in Unit IIa during the excavation of a 1 x 1-m square, S1W3. This excavation square is adjacent to the western wall of the cave and approximately 10 m beneath the surface. A flaked-stone projectile point was discovered in the upper-most portion of Unit IIc at a depth of 27 cm below the Holocene-Pleistocene boundary, and 25 cm below the split-bone point at a horizontal distance of 70 cm. Wood charcoal is abundant in the culture-bearing Unit IIa and IIb, and concentrated as a lens in the upper-most portion of Unit IIc.

For cross-dating, the flaked-stone and split-bone projectile points are the most reliable Paleoindian artifacts. The bone point has a beveled-base and cross-hachured surface. It was manufactured from a long, thick portion of megamammal cortical bone while it was still fresh and resilient. Stylistically, it is comparable to split-bone and ivory objects recovered from Paleoindian sites in Alaska, Saskatchewan, Oregon, Washington, Wyoming, Arizona, New Mexico, and Florida (Cressman 1942; Dunbar 1991; Dunbar and Webb 1996; Gramly 1993; Haynes 1982; Jenks and Simpson 1941; Rainey 1940; Simpson 1948; Wilmeth 1968; Yesner 1994).

Table 9.3. Stratigraphic contexts of diagnostic Paleoindian artifacts from Sheriden Cave

Unit	Diagnostic artifact	Charcoal
Ia	None	Scattered
Ib	None	Scattered
IIa	Beveled split-bone point	Abundant
IIb	None	Abundant
IIc	Fluted projectile point[a]	Lens[a]
IIIa	None	None
IIIb	None	None
IIIc	None	None
IVa	None	None
IVb	None	None
IVc	None	None
IVd	None	None
IVe	None	None

a = from upper-most portion of the unit

The complete and heavily reworked flaked-stone projectile point measures 3.62 cm in length, 1.65 cm in width, and 4.5 mm in thickness. Stylistically, the biface is similar to specimens recovered from the Holcombe Beach site in Michigan (Fitting et al. 1966). Sites with Holcombe points have never been radiocarbon dated. On the basis of beach ridges, their age estimations range from more than 11,000 to 10,000 radiocarbon yr B.P. (Ellis et al. 1998:154; Fitting et al. 1966:100).

The destruction and removal of sediments during the commercial expansion of the cave system in 1990 prevented investigations of residential space use or general location of activity areas in and around the site. Sheriden Cave is, however, geologically similar to sinkhole cave sites that

have been investigated in northern Florida (Dunbar 1991; Dunbar and Waller 1992; Dunbar and Webb 1996). It is quite possible that Paleoindian activity at Sheriden Cave was comparable to that in northern Florida during the late Pleistocene. In addition to split-bone and flaked-stone weaponry, other artifacts recovered from the cave suggest that animal resources were procured and processed at the site. They include burned and calcined bone, a large side scraper manufactured from nonlocal Wyandotte chert, a biface fragment of nonlocal Upper Mercer chert, a compass graver and a biface fragment manufactured from nonlocal Flint Ridge chert, the distal portion of an end scraper manufactured from local Pipe Creek chert, and a mixture of local (i.e., within 60 km) and nonlocal (i.e., between 140 and 400 km) debitage including microflakes and shatter of Wyandotte, Upper Mercer, and Flint Ridge cherts (Tankersley 1989:274-292). The debitage was likely produced from tool edge damage, resharpening dulled edges, and tool manufacture.

Chronometry

Chronometry is essential to fully understand the contextual environment and to integrate paleoecological data (Albanese 1996; Frison 1996; Julig et al. 1990; Julig and McAdrews 1993). Chronometric approaches are crucial because dynamic changes occur in the stratigraphic record. These changes are complex and result from both the original environmental context and the effects of the post-depositional environment (Kreutzer 1996).

Radiocarbon

Twenty-seven radiocarbon age determinations were obtained for the early Holocene and late Pleistocene strata (Table 9.4). Radiocarbon was atomically counted using accelerator mass spectrometry (AMS) on 11 samples of bone and dentin collagen from 5 mammals and 16 wood charcoal samples. A single conventional decay date was obtained on wood charcoal.

Table 9.4. Early Holocene and late Pleistocene radiocarbon dates from Sheriden Cave

Unit	Uncal. 14C yrs BP	Sample composition	Lab no.
Ia			
	No Dates		
Ib			
	9170 ± 60	Wood Charcoal	CAMS-24126
	9190 ± 60	Wood Charcoal	CAMS-24127
	9775 ± 70	Wood Charcoal	AA-21705
	10020 ± 115	Wood Charcoal	AA-21706
IIa			
	10550 ± 70	Wood Charcoal	Beta-117604
	10570 ± 70	Wood Charcoal	Beta-117605
	10600 ± 60	Wood Charcoal	Beta-117603
	10620 ± 70	Wood Charcoal	Beta-117606
	10680 ± 80	Wood Charcoal	AA-21710
	10850 ± 70	Wood Charcoal	Beta-117602
	10850 ± 60	Bone Collagen	CAMS-26783[a]
	10940 ± 70	Wood Charcoal	Beta-117601
	10970 ± 70	Wood Charcoal	Beta-117607

IIb			
	10470 ± 70	Wood Charcoal	AA-21712
	11060 ± 60	Bone Collagen	CAMS-10349[b]
	11130 ± 60	Bone Collagen	CAMS-33970[b]
	11710 ± 220	Wood Charcoal	PITT-0892
	13120 ± 80	Wood Charcoal	AA-21711
IIc			
	10840 ± 80	Wood Charcoal	Beta-127909
	10960 ± 60	Wood Charcoal	Beta-127910
	11480 ± 60	Bone Collagen	CAMS-12837[c]
	11570 ± 70	Bone Collagen	CAMS-12839[c]
	11570 ± 50	Bone Collagen	CAMS-33968[c]
	11610 ± 90	Bone Collagen	CAMS-12845[c]
	12520 ± 170	Dentin Collagen	Beta-127907[d]
	12590 ± 450	Dentin Collagen	Beta-127908a[d]
	12840 ± 100	Dentin Collagen	Beta-127908b[d]
IIIa			
	No Dates		
IIIb			
	No Dates		
IIIc			
	11530 ± 50	Dentin Collagen	Beta-127911[e]
IVa			
	No Dates		
IVb			
	No Dates		
IVc			
	No Dates		
IVd			
	No Dates		
IVe			
	No Dates		

a = *Castoroides ohioensis*
b = *Platygonus compressus*
c = *Arctodus simus*
d = *Cervalces scotti*
e = *Synaptomys borealis*

Four radiocarbon dates were obtained on wood charcoal from Unit Ib. They represent at least two populations of early Holocene dates: ca. 9,200 yr B.P. and ca. 9,900 yr B.P. Nine radiocarbon dates were obtained from Unit IIa, eight on wood charcoal and one on bone collagen. They depict two distinct populations of late Pleistocene dates: ca. 10,600 yr B.P. and ca. 10,900 yr B.P.

Four radiocarbon dates were obtained on wood charcoal and bone collagen from Unit IIb. They delineate three distinct populations of late Pleistocene dates: ca. 10,500 yr B.P., ca. 11,200 yr. B.P., and ca. 13,100 yr B.P. Post-depositional deformation features from loading in Unit IIb suggest that the three populations of dates represent a mixture of older and younger organic material from the overlying Unit IIa and underlying Unit IIc. The ca. 11,200 yr B.P. average of three dates from Unit IIb is perhaps the best age estimation for this stratum.

Nine radiocarbon dates were obtained on wood charcoal and bone collagen from Unit IIc. They characterize three distinct populations of late Pleistocene dates: ca. 10,900 yr B.P., ca. 11,600 yr B.P., and ca. 12,700 yr B.P. A single radiocarbon date was obtained for Unit IIIc on dentin collagen from the Northern Bog Lemming. A younger than expected assay of ca. 11,500 yr B.P. suggests that the date is better associated with burrowing activity than it is with a depositional episode.

Fluoride
Accurate chronologies are essential to demonstrating contemporaneity in non-analog late Pleistocene vertebrate communities (Semken et al. 1998). Radiocarbon, in specific non-degraded amino acids of collagen, atomically counted with accelerator mass spectrometry (AMS) is the single best method to directly date bone (Stafford et al. 1991). However, even high-precision AMS radiocarbon dates are reported with a one-sigma statistical error, that is, a minimum age estimation. Given that a two-sigma error provides a more probable assay of time, patterns in the synchroneity of certain species may simply be a by-product of time averaging, a significant limitation of our best dating method.

An important aspect of any comprehensive approach to chronological problems is age verification by a combination of dating techniques (Haynes 1992; Taylor et al. 1996). A more precise chronology can be made by plotting two independent measures of time. For almost 200 years, paleontologists have used fluoride content analysis as a method to provide a relative geological age of vertebrate fossils (Cook 1960; Middleton 1844). Today, fluoride concentrations can be measured in bone, dentin, or enamel using an Ion Selective Electrode (ISE).

By comparing ISE fluoride content analysis with the results of direct high-precision AMS radiocarbon dating of non-degraded amino acids, we can better evaluate the age of late Pleistocene species. For such comparisons, it is crucial that all radiocarbon dates are first calibrated because fluctuations and plateaus in atmospheric ^{14}C during the late Pleistocene make radiocarbon years appear compressed, representing less time in calendrical years (Taylor et al. 1996). Ideally, the curve of the calibrated radiocarbon dates should be linear with the logarithm of the concentration of fluoride.

For this study, AMS radiocarbon dates were obtained on the bone collagen of three extinct species: 10,850±60 BP (CAMS-26783) on *Castoroides ohioensis*; 11,130±60 BP (CAMS-33970) on *Platygonus compressus*; and 12,692±83 BP (average of Beta-127907, 127908a, 127908b) on *Cervalces scotti* (Semken et al 1998). ISE fluoride age determinations were also obtained for these individuals following the procedures outlined by Schurr (1989) and Tankersley et al. (1998). Fluoride content was measured in solution at the parts-per-million (ppm) and reported here in percentage. A linear relationship is illustrated between the log of the percent fluoride concentration and the calibrated radiocarbon dates. Given that the slope is time dependent, it may provide a calibration curve for future fluoride dates obtained on other species from this cave.

Twenty-seven fluoride age determinations were also obtained on post-cranial (i.e., tibia, ulna, metacarpal, femur, scapula, humerus, and innominate) flat-headed peccary bone from Unit II. All of the peccary bone samples produced fluoride solutions in excess of 20 ppm. With relative errors between 0% and 2% of the fluoride concentrations, determinations ranged from 0.4% to 2.2%. Eighty additional fluoride age determinations were obtained on the molars of five species of small rodents from Unit II. Of the 80 molars, the samples include muskrat *(Ondatra zibethica)*, yellow-cheeked vole, northern bog lemming, heather vole, and meadow vole *(Microtus pennsylvanicus)*. Unlike the peccary bone, the rodent molars produced fluoride solutions as low as 0 ppm and as high as 28 ppm. Fluoride determinations ranged from 0% to 1.3%. All of the fluoride determinations less than 0.4% were obtained on meadow voles. This species is still living in the cave today suggesting that their presence in Unit II is in part the result of late Holocene and modern burrowing activity. Interestingly, there is a remarkable consistency between the cumulative frequency curves for the radiocarbon and fluoride dates suggesting that sediment continuously accumulated in the cave during the late Pleistocene.

Magnetic Susceptibility
The magnetic susceptibility of Units IIa, IIb, and IIc were measured as another independent measurement of time (Figure 9.1). Variations in magnetic susceptibility are attributed to variations in climate controlled pedogenesis and the production of magnetic mineral phases outside of the cave. As surface soils accumulate in the cave from physical, biological, or cultural processes, they create magnetic susceptibility changes observable in the sediments (Ellwood et al. 1997). The low magnetic susceptibility magnitudes of Unit IIa and IIc indicate times of a cooler-dryer climate resulting from reduced pedogenesis. Unit IIb yielded higher magnetic susceptibility magnetudes indicative of warmer-wetter times from higher pedogenetic rates. If the radiocarbon ages for Units IIa, IIb, and IIc are calibrated to calendrical time, then the variations in magnetic susceptibility can be directly compared to the detailed stable-oxygen isotope ($d^{18}O$) record from the Greenland Ice-core Project (Figure 9.2) (Dansgaard et al. 1993). Interestingly, both records provide evidence for general climate instability between 12,000 and 16,000 years ago and illustrate a disctinctive paleoclimatic downturn or fingerprint for the Younger Dryas.

Discussion
The archaeology of Sheriden Cave, as well as most of the vertebrate and paleobotanical taxa, can be correlated with the late Pleistocene Younger Dryas chronozone. High-resolution stratigraphic, chronometric, and paleontological data suggest that the Younger Dryas was locally represented as a period of climatic osscilation and extremely fast environmental change. It correlates with a period of megamammal extinction and a marked decrease in $d^{18}O$ records (Haynes 1991).

The composition of vertebrate species and their abundance in the late Pleistocene and early Holocene units significantly and continuously changed. These changes may be the result of individual responses of the biota creating new community patterns and, perhaps, extinction of the megamammals (Graham and Grimm 1990; Webb et al. 1993; Semken et al.

1998). The issue of megamammal extinction is of more than a little archaeological interest because it would have influenced the course of human movement and economic divergence (Tankersley and Isaac 1990). Demonstrating the cause of late Pleistocene extinctions remains a hotly debated issue. Overkill, climatic change, and some combination of these endure as the major competing causal hypotheses (Fisher 1996).

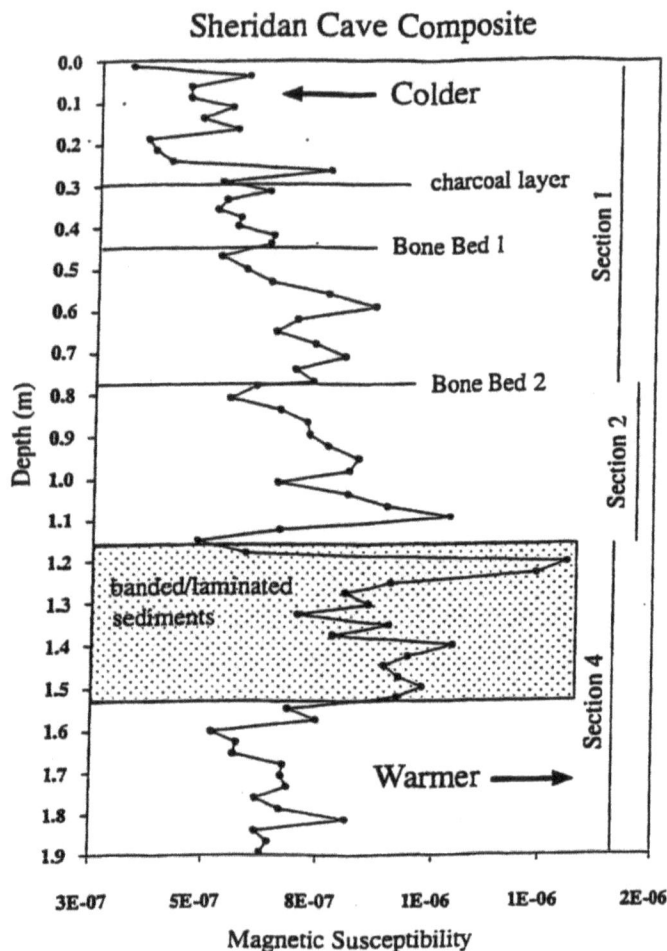

Figure 9.1. High-resolution paleoclimatic trends for magnetic susceptibility data from Unit II.

Direct AMS radiocarbon dating of collagen demonstrate that the disappearance of megamammals occurred over a period of approximately 2,000 radiocarbon years, between ca. 12,700 and 10,900 yr B.P. Chronometric dates also suggest that extirpations or extinctions were not simultaneous. Megamammals do not disappear from the stratigraphic record at the same time. To some extent, their presence and size seem to shift through time--from older and larger to smaller and more recent species. For example, the stag-moose dates ca. 12,700 radiocarbon yr B.P., the short-faced bear ca. 11,600 radiocarbon yr B.P., the flat-headed peccary ca. 11,100 radiocarbon yr B.P., and the giant beaver ca. 10,900 yr B.P. Similar patterns have been described for late Pleistocene deposits on the Great Plains of North America

72

and eastern Siberia (Frison 1996; Frison and Bonnichsen 1996; Tankersley and Kuzmin 1998).

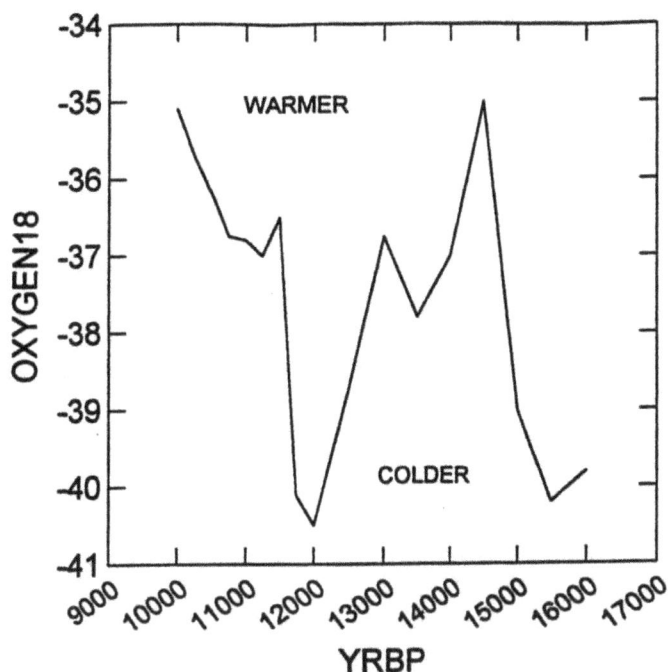

Figure 9.2. Oxygen stable-isotope data from the Greenland Ice-core Project (after Dansgaard et al. 1993)

The change in the composition of non-analog species in the late Pleistocene deposits is penecontemporaneous with the disappearance of megamammals suggesting that climate rather than overkill was at least a contributing cause of their extinction (Semken et al. 1998). In this scenario, megamammal extinction would have been highly variable and related to the diversity of the surviving plant and animal resources and their individual response to climatic change (Yesner 1996).

On the other hand, the disappeance of megamammals is approximately coincident with at least one Paleoindian component from Sheriden Cave. The culture-bearing deposits date within 400 radiocarbon years, between ca. 10,600 and 10,900 yr B.P. This radiocarbon age range coincides with the recently dated extinction of at least 30 species of megamammals in North America (Semken et al. 1998). This overlap is further evidenced by the fact that the split-bone point was manufactured from megamammal cortical bone while it was in a fresh state.

The presence of a Holcombe projectile point in the late Pleistocene deposits may also indicate that Paleoindian co-traditions, or at least multiple technological complexes, were present in eastern North America during the late Pleistocene. However, better temporal resolution from well-stratified sites that include the late Pleistocene and early Holocene boundary will be required to ultimately resolve this issue.

Acknowledgments. Field and laboratory research consists of an interdisciplinary team including Robert Brackenridge (Dartmouth College, geologist), Brooks Ellwood (University of Texas, geophysicist), Frances King (Cleveland Museum of Natural History, paleobotanist), Patrick Munson (Indiana University, geoarchaeologist), Brian Redmond (Cleveland Museum of Natural History, archaeologist), Donald Stierman (University of Toledo, geophysicist), and Greg McDonald (National Park Service, paleontologist). Research for this essay was supported by a grant from the National Science Foundation (SBR 9707984). This project would not be possible without the help and cooperation of Jean Hendricks, Keith Hendricks, Danny Tong, the Cincinnati Museum of Natural History, and the efforts of scores of devoted volunteer field workers. The long-term contributions of Cheryl Birkhimer, Philip Cossentino, Paul Barans, Elaine Dowd, Lucinda McWheeny, Larry and Nancy Morris, Carl Syfranski, Garry Summers, and Jenny Tankersley were invaluable. This essay is dedicated to the memory of Richard "Dick" Hendricks, without whom none of this research would have been possible.

References

Albanese, J., 1996. Geology of the Mill Iron Site. In *The Mill Iron Site*, ed. G. Frison. Albuquerque: University of New Mexico Press, pp. 25-42.

Anderson, T.W. and C.F.M. Lewis, 1985. Postglacial water-level history of the Lake Ontario Basin. In *Glacial Lakes in the Ontario Basin*, ed. E. Muller and V. Prest. St. John's: Geological Association of Canada, pp. 231-253.

Barnett, P.J., 1985. Glacial retreat and lake levels, north-central Lake Erie Basin. In *Glacial Lakes in the Ontario Basin*, ed. E. Muller and V. Prest. St. John's: Geological Association of Canada, pp. 185-194.

Calkin, P.E., and B.H. Feenstra, 1985. Evolution of the Lake Erie Basin Great Lakes. In *Glacial Lakes in the Ontario Basin*, ed. E. Muller and V. Prest. St. John's: Geological Association of Canada, pp. 149-170.

Carlson, E. H., 1991. *Minerals of Ohio*. Columbus: Ohio Geological Survey, Bulletin 69.

Cook, S.F., 1960. Dating prehistoric bone by chemical analysis. In *The Application of Quantitative Methods in Archaeology*, ed. R. Heizer and S. Cook. New York: Wenner-Gren Foundation, pp. 223-245.

Cressman, L., 1942. *Archaeological Researches in the Northern Great Basin*. Carnegie Institution of Washington Publication, No. 58.

Dansgaard, W., S. J. Johnson, H. B. Clausen, D. Dahl-Jensen, N. S. Gunderstrup, C. U. Hammer, C. S. Hvidberg, J. P. Steffensen, A. E. Sveinbjornsdottir, J. Jouzel, and G. Bond, 1993. Evidence for General Instability of Past Climate from a 250-kyr Ice Core Record. *Nature* 364, 218-220.

Dunbar, J. S., 1991. Resource Orientation of Clovis and Suwannee Age Paleoindian Sites in Florida. In *Clovis Origins and Adaptations*, ed.R. Bonnichsen and K. L Turnmire. Corvallis: Peopling of the Americas Publications, pp. 185-213.

Dunbar, J. S., S. D. Webb, M. Faught, 1991. Inundated prehistoric sites in Apalachee Bay, Florida, and the search from the Clovis shoreline. In *Paleoshorelines and Prehistory*, ed. L. Johnson and M. Stright. Boca Raton, Florida: CRC Press, pp. 117-146.

Dunbar, J.S., and B.I. Waller, 1992. Resource orientation of Clovis, Suwannee, and Simpson age Paleoindian sites in Florida. In *Paleoindian and Early Archaic Period Research in the Lower Southeast: A South Carolina Perspective*, ed. D.G. Anderson, K.E. Sassaman, and C. Judge. Columbia: Council of South Carolina Professional Archaeologists, pp. 279-295.

Dunbar, J.S., and S.D. Webb, 1996. Bone and ivory tools from Paleoindian sites in Florida. In *The Paleoindian and Early Archaic Southeast*, ed. D.G. Anderson and K.E. Sassaman. Tuscaloosa: University of Alabama Press, pp. 331-353.

Ellis, C., A.C. Goodyear, D.F. Morse, and K.B. Tankersley, 1998. Archaeology of the Pleistocene-Holocene transition in eastern North America. *Quaternary International* 49/50, 151-166.

Ellwood, B. B., K. M. Petruso, and F. B. Harrold, 1997. High-Resolution paleoclimatic trends for the Holocene identified using magnetic susceptibility data from archaeological excavations in caves. *Journal of Archaeological Science* 24, 569-573.

Fisher, D. C., 1996. Extinction of proboscideans in North America. In *Evolution and Paleoecology of Elephants and their Relatives*, ed. Shoshan and Tassy. Oxford: Oxford University Press, pp. 296-315.

Fitting, J.E., J. Devisscher, and E.J. Wahla, 1966. *The Paleoindian Occupation of the Holcombe Beach*. Ann Arbor: Anthropological Papers, No. 27, Museum of Anthropology, The University of Michigan.

Ford, K.M., A.R. Bair, and J.A. Holman, 1996. Late Pleistocene fishes from Sheriden Pit, northwestern Ohio. *Michigan Academician* 28, 135-145.

Frison, G. C., 1996. Introduction. In*The Mill Iron Site*, ed. G. C. Frison. Albuquerque: University of New Mexico Press, pp. 205-216.

Frison, G.C., and R. Bonnichsen, 1996. The Pleistocene-Holocene transition on the Plains and Rocky Mountains of North America. In *Humans at the End of the Ice Age: The Archaeology of the Pleistocene-Holocene Transition*, ed. L. Straus, B. Eriksen, J. Erlandson, and D. Yesner. New York: Plenum Press, pp. 303-318.

Gobetz, K. E., 1998. Morphological and Taphonomic Study of a Local Assemblage of Flat-Headed Peccaries (*Platygonus compressus* Leconte) from a Late Pleistocene Sinkhole Deposit in Northwestern Ohio, U.S.A. Unpublished M.S. Thesis, Department of Geology, Indiana University, Bloomington.

Graham, R. W., and E. C. Grimm, 1990. Effects of global climate change on the patterns of terrestrial biological communities. *Trends in Ecology and Evolution* 5, 289-292.

Gramly, R.M., 1993. *The Richey Clovis Cache*. Buffalo, New York: Persimmon Press.

Haynes, C. V., 1982. Were Clovis progenitors in Beringia? In *Paleoecology of Beringia*, ed. D.M. Hopkins, J.V., Matthews, C.E. Schweger and S.B. Young. New York: Academic Press, pp. 383-398.

Haynes, C. V., 1991. Geoarchaeological and paleohydrological evidence for a Clovis-age drought in North America and its bearing on extinction. *Quaternary Research* 35, 438-450.

Haynes, C. V., 1992. Contributions of radiocarbon dating to the geochronology of the peopling of the New World. In *Radiocarbon Dating After Four Decades*, ed. R.E. Taylor, A. Long, and R.S. Kra. New York: Springer-Verlag, pp. 355-374.

Holman, J.A., 1997. Amphibians and reptiles from the Pleistocene (Late Wisconsinan) of Sheriden Pit Cave, northwestern Ohio. *Michigan Academician* 29, 1-20.

House, V.H., 1985. Pleistocene Geology of Wyandot County, Ohio. Unpublished M.S. thesis, Department of Geology, Bowling Green State University, Ohio.

Hoy, R.G., J.M. Harbor, and E.H. Carlson, 1995. The origin of fine-grained sediment in the Ohio caverns. *Northeastern Geology and Environmental Sciences* 17, 83-88.

Jenks, A.E., and H.H. Simpson, 1941. Beveled artifacts in Florida of the same type as artifacts found near Clovis, New Mexico. *American Antiquity* 6, 314-319.

Julig, P. J., and J. H. McAndrews, 1993. Paleoindian cultures in the Great Lakes region of North America paleoclimatic, geomorphic, and stratigraphic context. *L'Anthropologie* 97, 623-650.

Julig, P . J., J. H. McAndrews, and W. C. Mahaney, 1990. Geoarchaeology of the Cummins Site on the beach of Proglacial Lake Minong, Lake Superior basin, Canada. In *Archaeological Geology of North America*, ed. N. P. Lasca and J. Donahue. Boulder: Geological Society of America, pp. 21-50.

Kreutzer, L. A., 1996 Summary of skeletal parts and X-ray densitometry measurements of bison bone material. In *The Mill Iron Site*, ed. G. Frison. Albuquerque: University of New Mexico Press, pp. 225-230.

McDonald, H.G., 1994. Late Pleistocene vertebrate fauna of Ohio: coinhabitants with Ohio's Paleoindians. In *The First Discovery of America*, ed. W. S. Dancey. Columbus, The Ohio Archaeological Council, pp. 23-42.

Middleton, J., 1844. On fluorine in bones, its source, and its application to the determination of the geological age of fossil bones. *Proceedings of the London Geological Society* 4, 431-433.

Rainey, F., 1940. Archaeological investigations in Central Alaska. *American Antiquity* 4, 299-308.

Schurr, M. R., 1989. Fluoride dating of prehistoric bone by ion selective electrode. *Journal of Archaeological Science* 16, 265-270.

Semken, H.A., T.W. Stafford, and R.W. Graham, 1998. Contemporaneity of megamammal extinctions and the reorganization of non-analog micromammal associations during the late Pleistocene of North America. *Final Program and Abstracts of the 8th*

International Congress of the International Council for Archaeozoology, 257.

Simpson, J.C., 1948. Folsom-like points from Florida. *Florida Anthropologist* 1,11-15.

Stafford, T. W., P. Hare, L. Curie, A. Jull, 1991. Accelerator radiocarbon dating at the molecular level. *Journal of Archaeological Science* 18, 35-72.

Tankersley, K.B., 1989. A close look at the big picture: early Paleoindian lithic procurement in the midwestern United states. In *Paleoindian Lithic Resource Use*, ed. C. Ellis and J. Lathrop. Boulder: Westview Press, pp. 259-292.

Tankersley, K.B., 1997. Sheriden: a Clovis site in eastern North America. *Geoarchaeology* 12, 713-724.

Tankersley, K.B., and B.L. Isaac, 1990. Concluding Remarks on paleoecology and paleoeconomy. In *Early Paleoindian Economies of Eastern North America*, ed. K. Tankersley and B. Isaac. Research in Economic Anthropology, Supplement 5. Greenwich: JAI Press, pp. 337-355.

Tankersley, K. B., and Y. V. Kuzmin, 1998. Patterns of culture change in Eastern Siberia during the Pleistocene-Holocene transition. *Quaternary International* 49/50, 129-140.

Tankersley, K.B., K.D. Schlecht, and R.S. Laub, 1998. Fluoride dating of Mastodon bone from an early Paleoindian spring site. *Journal of Archaeological Science* 25, 1-7

Taylor, R. E., C. V. Haynes, and M. Stuiver, 1996. Clovis and Folsom age estimates: stratigraphic context and radiocarbon calibration. *Antiquity* 70, 515-525.

Totten, S.M., 1985. Chronology and nature of the Pleistocene beaches and wave-cut cliffs and terraces, Northeastern Ohio. In *Glacial Lakes in the Ontario Basin*, ed. E. Muller and V. Prest. St. John's: Geological Association of Canada, pp. 171-184.

Webb, T., P. J. Bartlein, S. P. Harrison, and K. H. Anderson, 1993. Vegetation, lake levels, and climate in Eastern North America for the past 18,000 years. In *Global Climates Since the Last Glacial Maximum*, ed. H. Wright, J. Kutzbach, T. Webb, W. Ruddiman, F. street-Perrot, and P. Bartlein. Minneapolis: University of Minnesota Press, pp. 415-465.

Wilmeth, R., 1968. A fossilized bone artifact from Southern Saskatchewan. *American Antiquity* 33, 100-101.

Yesner, D. R., 1994. Subsistence diversity and hunter-gatherer strategies in Late Pleistocene/Early Holocene Beringia: evidence from the Broken Mammoth Site, Big Delta, Alaska *Current Research in the Pleistocene* 11,154-156.

Yesner, D. R., 1996. Human adaptation at the Pleistocene-Holocene boundary (ca. 13,000 to 8,000 yr BP) in Eastern Beringia. In *Humans at the End of the Ice Age: The Archaeology of the Pleistocene-Holocene Transition*, ed. L. Straus, B. Eriksen, J. Erlandson, and D. Yesner. New York: Plenum Press, pp. 255-276.

STRATIFIED FAUNAS FROM CHARLIE LAKE CAVE AND THE PEOPLING OF THE WESTERN INTERIOR OF CANADA

Jonathan C. Driver

Department of Archaeology, Simon Fraser University, Burnaby, BC V5A 1S6, Canada

Introduction

At the end of the Pleistocene in many parts of the world there were significant and rapid environmental changes. For people in those regions, environmental change could have had three effects. First, some locations became uninhabitable, as was the case in many coastal regions inundated by rising sea levels. Second, in order to maintain populations in areas which did not become uninhabitable, new subsistence strategies were required; changing patterns of foraging, new settlement systems, or the domestication of plants and animals are examples of these strategies. Third, areas which had previously been uninhabitable or inaccesible were open for colonization. This paper deals with the latter topic by considering the evidence for early post-glacial environments in western Canada and the use of animal resources during the early post-glacial period.

Much of the discussion about early human cultures in the Americas has concerned the timing of the initial colonization, and especially the route and chronology for movement of people from northeast Asia through Alaska and western Canada and south into unglaciated territory from the Canada/U.S.A. border to the southern tip of South America. Even if one argues for a relatively late entry of people to the Americas (for example, seeing the Clovis culture at about 11,500 B.P. as the earliest manifestation of human presence), it is certain that people lived both north and south of the Laurentide and Cordilleran ice sheets which covered most of what is now Canada. The melting of these ice sheets therefore opened up an area of more than 10 million square kilometres for colonization in the early post-glacial period.

Data from western Canada summarized for the meeting of the INQUA Working Group on the Archaeology of the Pleistocene-Holocene Transition in Berlin (Driver 1998a) have not changed significantly in the past few years. At 12,000 BP most of western Canada was covered by ice or glacial lakes, with coastal refugia providing a possible route for late Pleistocene entry to land south of the ice sheets, if that had not already occurred prior to the last extensive glaciation. By 11,500 BP plant and animal fossils show the re-establishment of biotic communities in southern Alberta, and by 10,000 BP most of western Canada south of 60° N was probably inhabitable. Unfortunately, the archaeological record prior to 10,000 BP is very sparse, and becomes even sparser when sites with faunal assemblages are considered. Only two published archaeological sites (Vermilion Lakes and Charlie Lake Cave) pre-date 10,000 BP and contain faunal assemblages with hundreds of specimens. In this paper I examine the fauna from Charlie Lake Cave with a view to assessing the adaptation of the first people to move into the recently deglaciated western interior of Canada.

Chronology and environmental setting

This paper is concerned with interior western Canada, which can be defined broadly as land lying east of the Rockies and west of the Canadian Shield. Because the western edge of this area was deglaciated first, attention is focused on the "western corridor" - a roughly 300 km wide strip of land to the east of the Rockies. Today this area is characterized by three major ecological zones - grassland, parkland and boreal forest - which succeed each other from south to north. The Rockies are mainly forested. First Nations who occupied these zones had conspicuously different adaptations. Bison (*Bison bison*) was the most important resource for inhabitants of grassland, whereas a variety of ungulates, especially moose (*Alces alces*) and caribou (*Rangifer tarandus*), fish, and waterfowl formed a more diverse subsistence base in the boreal forest. Plants were consumed in all areas, but were not dietary staples.

The radiocarbon chronology for deglaciation and the re-establishment of inhabitable biotic environments in western Canada remains contentious for two reasons:
1. Considering the size of the region, there are relatively few radiocarbon dates and a dearth of early archaeological, palynological and paleontological sites.
2. Before the development of AMS dating, bulk samples of sediments were sometimes used for dating, especially of pollen cores. Such samples are often contaminated by organics eroded from more ancient deposits, resulting in erroneously old radiocarbon dates. In addition, some aquatic plants obtain ancient carbon from dissolved bicarbonates, and these also provide dates which are too early (Beaudoin 1993: MacDonald et al. 1987, 1991; Wilson 1993).

If one is cautious about accepting radiocarbon dates on bulk organics derived from lake bed sediments, then the earliest post-glacial vegetation east of the Rockies probably appeared around 11,500 BP (MacDonald and McLeod 1996). Mandryk (1996) has pointed out that vegetation may have developed on stagnant ice. If such vegetation did exist, it may pre-date the formation of modern lakes and bogs which now hold the oldest palynological records. However, as will be seen, the evidence from vertebrate fossils shows no large mammals in most of the western corridor before 11,500 BP, so if there was a vegetation cover over stagnant ice it probably could not support human populations.

Palynological data show that deglaciated landforms were typically colonized by a vegetation cover which has no modern analogues (Lichti-Federovich 1970; MacDonald 1987; MacDonald and McLeod 1996; White and Mathewes 1986). The vegetation consisted mainly of grasses, sedges, herbs and shrubs. *Populus* sp. (probably aspen) is the most common tree represented, and may have been more common than suggested by pollen frequencies due to poor preservation of its pollen. The vegetation seems to have consisted of a mix of species which could colonize poor soils relatively rapidly. It may have resembled a combination of steppe/grassland, parkland and wet tundra. Pollen deposition rates are similar to those found on modern grassland margins (MacDonald and McLeod 1996). Except for areas which remain as grassland today, much of the western corridor was subsequently colonized by coniferous taxa, notably spruce, at about 10,000 BP. The development of this early boreal forest effectively marks the end of the early post-glacial vegetation.

Vertebrate fossils from the region provide a less detailed picture of the environments and environmental change, but do provide evidence for the absence of inhabitable landscapes during late glacial times. Burns (1996) has shown that dated vertebrates are either older than about 20,000 BP (i.e. they pre-date the last glaciation) or younger than 12,000 BP (i.e. they post-date the last glaciation). The absence of vertebrates dating between about 20,000 and 12,000 BP strongly suggests that western Canada could not support large mammals during the last glaciation, and that human occupation is also unlikely. Further evidence for the lack of inhabitable environments in the interior of western Canada comes from the species composition in early post-glacial times. With the exception of the most southerly areas, early post-glacial faunas are dominated by *Bison*. Typical North American late glacial species from south of the ice sheets, such as mammoth, horse and camel have not been found over most of the region in early post-glacial deposits, suggesting that large fauna did not colonize the area until after most of the late Pleistocene extinctions had taken effect. Given the common appearance of *Bison* (Wilson 1996), it would be remarkable if other taxa had somehow been missed by paleontologists, especially as they have been recovered from earlier deposits. Furthermore, early post-glacial bison in interior western Canada are most similar to specimens further south, and are different from specimens living north of the ice sheets (Wilson 1996).

Thus a cautious interpretation of the early post-glacial environmental sequence would see a relatively late deglaciation (c. 12,000 BP), followed by about 2000 radiocarbon years of open landscapes. During this time animal species moved into the western corridor from the south, with *Bison* (a survivor of late Pleistocene extinctions) the dominant large animal. Over much of the interior of western Canada the open landscapes were replaced by spruce-dominated boreal forest at about 10,000 BP.

Human history
I have reviewed western Canadian archaeological data elsewhere (Driver 1998a), and there has been little significant change since that review. Using a recently established database of Canadian radiocarbon dates (Morlan 1999) one can examine large scale patterns in dated sites (Figure 10.1). Canadian archaeological sites are given "Borden numbers" (combinations of letters and digits) according to their geographic location (Borden 1952). I have selected for analysis five "Borden blocks", running roughly northwest along the east foothills of the Rockies and adjacent plains. This area has yielded the greatest number of pre-9000 BP sites in western Canada. Each Borden block is identified by two letters (e.g. EP), and each encompasses two degrees of latitude and four degrees of longitude, except DP whose lower half is truncated by the Canada/U.S.A. border.

Figure 10.1 plots the number of sites dated to 1000 radiocarbon year intervals for five Borden blocks running south to north. A site was only included in a 1000 year interval once - i.e. multiple dates or multiple components from the same millenium and the same site were ignored. However, sites with dates from different 1000 year intervals were counted separately for each millenium. The small number of dated sites in all periods in the more northern Borden blocks is probably the result of low-intensity archaeological fieldwork, coupled with low sedimentation rates and poor preservation. For example, most of the dates for the HR block are from Charlie Lake Cave, discussed below. The relatively sharp increase in site numbers from 11,000 to 9000 BP is probably the result of (a) increasing prehistoric populations, and (b) increasingly stable land surfaces following a great deal of early post-glacial landscape remodelling. The earliest post-glacial landscapes were often either eroded or buried deeply (Ryder 1971), which means that it is difficult to find early sites, especially as accessible limestone caves are rare. Most of the radiocarbon dated pre-9000 BP sites listed in a previous study (Driver 1998a: Table 1) are buried by at least a metre of sediment, and some are much more deeply buried. Most were discovered either when excavating below a later prehistoric component or as a result of development activity.

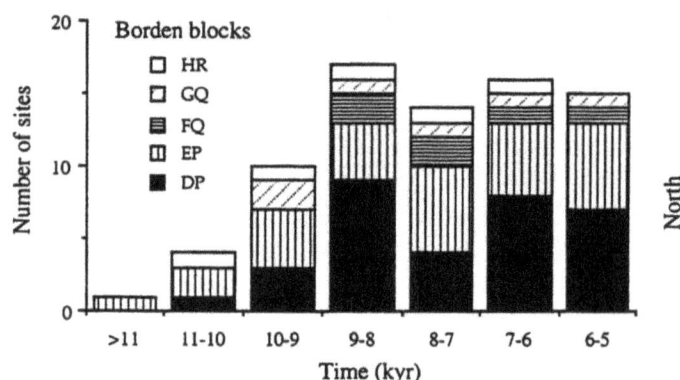

Figure 10.1. Number of dated archaeological sites or components per 1000 radiocarbon year intervals by Borden blocks.

As the major modern environmental patterns were established in western Canada by 9000 BP, one can see from Figure 10.1 that very few sites are available for analysis of the Pleistocene-Holocene transition. Most of the sites from the early period either lack fauna, or bones occur in small numbers. It is therefore impossible to detect temporal or spatial patterns in animal distribution. The remainder of this paper focuses on one site - Charlie Lake Cave - where stratified, dated faunal assemblages are associated with human occupations.

Charlie Lake Cave
Chronology
Charlie Lake Cave is located just north of the Peace River valley, about 160 km east of the continental divide at 56°16'35"N, 120°56'15"W, 730 m asl. The site, its stratigraphy and archaeological sequence have been described elsewhere (Fladmark et al. 1988; Driver et al. 1996, and

references therein). The major features of the site are as follows.

1. Most materials have been recovered from roughly 4m of sediment which fills a gully in front of the cave entrance.

2. The sequence of deposits dates from about 10,500 BP to the present.

3. Archaeological components are found intermittently through the sequence, but there is a notable gap from about 9500 to 7000 BP with no human use of the site.

4. Faunal remains are found throughout the sequence, regardless of presence or absence of people.

Most of the deposits consist of a mixture of glacial lake silts redeposited from the hillside above the site and weathered sandstone from the local bedrock. Palaeosols have formed on these sediments from time to time, and a combination of granulometry and palaeosol formation has been used to subdivide the stratigraphy into zones and subzones. This paper uses the most recent stratigraphic nomenclature (see Driver et al. 1996). When looking at the Pleistocene/Holocene boundary we are concerned mainly with Zone II, which represents rapid redeposition of glacial lake sediments, and the lower part of Zone III in which sedimentation rates slowed, more weathered sandstone occurs, and the first palaeosols appear.

Zone II has been subdivided into four subzones. Subzones IIa and IIb are the earliest deposits which contain bones and artifacts, and six radiocarbon dates average 10,500 ± 40 BP. Subzones IIc and IId are very similar. Four radiocarbon dates, all from IIc, average 9850 ± 80 BP. Subzone IIIa is dated by a single date of 9490 ± 140 BP. Two dates from subzone IIIb average 8350 ± 230 BP. All subzones except IIIb contain stone artifacts. Details of artifacts and dates can be found in Handly (1993) and Driver et al. (1996). The sequence of radiocarbon dates agrees well with the stratigraphic sequence (Figure 10.2), and the rapid sedimentation in Zone II (up to one metre of sediment in 1000 radiocarbon years) has probably reduced the chance of mixed assemblages.

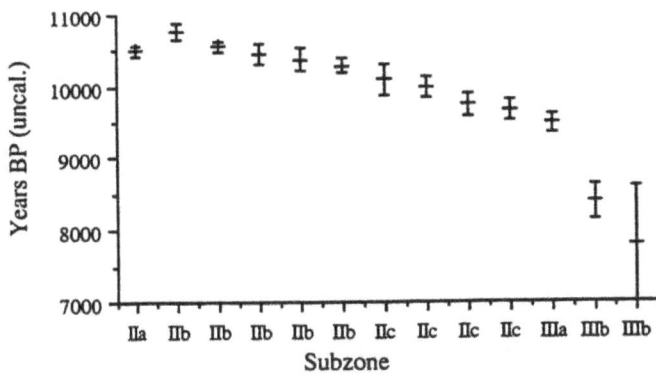

Figure 10.2. Radiocarbon dates from Zone II and lower Zone III, Charlie Lake Cave. Dates are organized by decreasing age within subzones.

Archaeology
Artifacts have been described in detail elsewhere (Handly 1993; Driver et al. 1996). With the exception of a drilled

stone bead, all artifacts are of flaked local stone, mainly cherts with some quartzite. The array of tool forms is consistent with hunting and processing of animals. Few artifacts have sufficient distinctive formal qualities to allow comparison with dated sites elsewhere, and there are very few dated sites within 500 km. A spear point from the lowest cultural component (subzone IIb) is related to fluted point complexes to the south. A microblade core from subzone IIIa has some similarities to wedge-shaped cores from Alaska, but is technologically different. Debitage suggests little on-site production of artifacts; instead, artifact maintenance is represented by small resharpening flakes, and discarded formed artifacts are relatively common. This, together with the absence of hearths or charcoal, suggests relatively short-term use of the site in early post-glacial times. This contrasts with later (post-5000 BP) occupations, where there is evidence for artifact manufacture and hearth features.

Identified taxa and palaeoenvironments
Table 1 lists mammalian and avian specimens which have been identified to genus or species, as well as occasional family level identifications (e.g. Rallidae) where the family level identification provides useful palaeoenvironmental evidence. Taxa are grouped by their likely habitat preferences; "catholic" taxa are today associated with a wide variety of habitats.

The earliest fauna, from Subzones IIa and IIb, is different from all later faunas in a number of ways. It contains taxa which prefer open habitats and which are no longer found in the Peace River region. The most notable species is *Dicrostonyx torquatus* (collared lemming), which had a wider late Pleistocene distribution, but is now found only on tundra (Driver 1998b). Ground squirrels (*Spermophilus*), bison and a large hare are also associated. Forest mammals are notably absent, with the exception of snowshoe hare (*Lepus americanus*), which is represented by only a few specimens. Equally notable is the absence of waterfowl and aquatic mammals, suggesting that stable, productive, aquatic environments had not been established near the site.

Table 10.1. Identified fauna from Charlie Lake Cave. See text for details of inclusion of taxa.

TAXON	IIa,IIb	IIc,IId	IIIa	IIIb
Open habitat mammals				
Lepus sp.	*	*		
Large hare				
Dicrostonyx torquatus	*			
Collared lemming				
Spermophilus sp.	**	*	*	
Ground squirrel				
Bison sp.	*	*	*	*
Bison				
Aquatic birds				
Aechmophorus sp.		*	*	*
Western or Clark's Grebe				

79

Podiceps auritus		*	*	
Horned grebe				
Anas platyrhynchos		*		
Mallard				
Anas sp.		*		*
Teal				
Bucephela albeola				*
Bufflehead				
Oxyura jamaicensis		*		*
Ruddy duck				
Fulica americana		*	*	
Coot				
Rallidae		*		
Virginia or Sora rail				
Aquatic mammals				
Ondatra zibethicus			*	*
Muskrat				
Castor canadensis			*	*
Beaver				
Forest birds				
Surnia ulula				*
Hawk owl				
Picidae	*			
Woodpecker				
Ectopistes migratorius			*	
Passenger Pigeon				
Forest mammals				
Lepus americanus	*	**	**	**
Snowshoe hare				
Microtus xanthognathus		*	*	
Chestnut-cheeked vole				
Clethrionomys gapperi		*	*	*
Gapper's red-backed vole				
Marmota monax				*
Woodchuck				
Catholic birds				
Tetraonidae		*	*	*
Grouse				
Asio flammeus		*		
Short-eared owl				
Corvus corax	*	*		
Raven				
Hirundo pyrrhonota	*	*	*	
Cliff swallow				

Catholic mammals				
Peromyscus sp.	*	*	*	*
Deer mouse				
Neotoma sp.			.	*
Packrat				
Eutamias sp.			*	*
Chipmunk				
Canis sp.		*		
Wolf-size canid				
Mustela nivalis			*	
Least weasel				
Mephitis mephitis			*	*
Striped skunk				

* = present   ** = common (NISP > 100)

Beginning in subzones IIc and IId, there is an increase in the number of taxa associated with aquatic environments and with the boreal forest. Snowshoe hare becomes common, and voles commonly associated with damp forest habitats occur. By the end of subzone IIIa the last of the ground squirrels disappear, and bison is the only grassland-adapted taxon present. Bison today occupy boreal forest in relatively low population densities; they probably persisted in the Peace River region as a result of the "Peace River grasslands" - areas of grassland and parkland within the southern boreal forest, mainly to the east of Charlie Lake Cave (White and Mathewes 1986: Figure 1). Thus by 9000 BP the fauna represented at Charlie Lake Cave is essentially modern, at least in terms of taxa represented.

The transition from relatively open to forested environments can be examined in more detail at Charlie Lake Cave because the relatively rapid sedimentation rates allow good separation of assemblages. The dominant small mammals of Zone II assemblages are ground squirrels and snowshoe hare. Figure 10.3 plots the relative frequency of these two taxa in three excavation units which are relatively deep and well dated. Units 26 and 28 are next to each other. Unit 3 is a metre away from them. Unit 3 was excavated in 1983 and units 26 and 28 were excavated in 1991, and different numbering systems were used for stratigraphic layers. Subzones IIb (layers 105, 104), IIc (layer 98), IId (layer 93) and IIIa (layers 82 through 91) are present. The data suggest that snowshoe hare appeared at about 10,000 BP, and became a common component of the local fauna within a few hundred radiocarbon years.

If we assume that ground squirrels were associated with relatively open, treeless environments, and that snowshoe hare was associated with coniferous forests, then Charlie Lake Cave provides one of the few post-glacial vertebrate records in western Canada which documents the transition from open to forested conditions. Nearby pollen records also document this transition. Of particular interest are dates and pollen zones from Lone Fox Lake to the northeast (MacDonald 1987) and Boone Lake to the southeast (White and Mathewes 1986). Dates for the local appearance of spruce at Boone Lake are about 10,700 BP. At Charlie Lake Cave and Lone Fox Lake this is a time of open vegetation.

Spruce forest was local around Lone Fox Lake by 9800 BP, which is consistent with the dates for the disappearance of ground squirrels and dominance of snowshoe hare at Charlie Lake Cave. As radiocarbon dates for the appearance of spruce are based on bulk sediment samples from Boone Lake, it is possible that the dates are somewhat older due to the "old carbon" effect. Alternatively, the more southern Boone Lake area may have been colonized by spruce earlier, with a subsequent lag in spruce migration north caused by a short cool period, possibly relating to the Younger Dryas. Possible evidence for a western interior cool episode was suggested for pollen data from the Vermilion Lakes site (Fedje et al. 1995: 102) and has been reported for the northwest coast (Mathewes 1994). Given the problems associated with radiocarbon dating at this time period, potential contamination of lake bed samples, problems of comparing plant and bone dates, and the variable sensitivity of vertebrates and plants to environmental change, there is remarkably good concordance between environmental reconstructions based on local pollen sequences and the Charlie Lake Cave vertebrates.

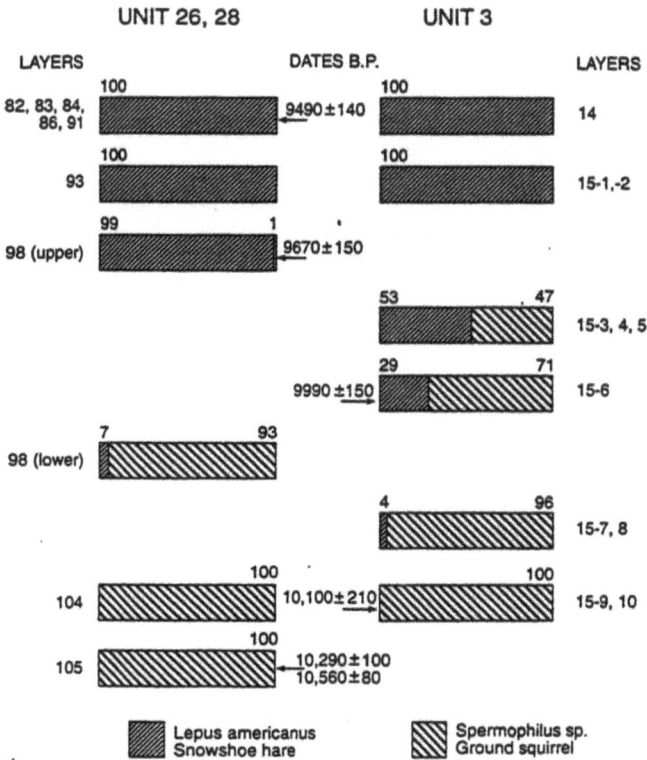

Figure 10.3. Relative frequency (%NISP) of ground squirrels and snowshoe hare for three excavation units.

Human adaptation
Taphonomic analysis of Charlie Lake Cave faunas is not yet complete, but it is clear that separating culturally deposited from naturally deposited specimens will be extremely difficult. Deposition of a wide variety of vertebrates occurred during times when the site was not visited by people, and one cannot assume that any taxon owes its presence on the site to human activity. Cutmarks are very rare. Burning is not present on any of the pre-9000 BP specimens, but is

common in mid- and late-Holocene assemblages. For the early post-glacial assemblages discussed in this paper only bison displays evidence of cutmarks and spiral fractures with impact points (Fladmark et al. 1988). As discussed elsewhere (Driver 1998a), large game animals are typically associated with late Pleistocene and early Holocene archaeological sites in the western interior of Canada, and hunting of large game is a logical adaptation to the early post-glacial open environments. Bison is the most frequently found large mammal, but at Vermilion Lakes bighorn sheep was dominant and caribou was probably present (Fedje et al. 1995).

The early post-glacial hunter-gatherers in this region probably migrated from the south, taking advantage of newly created habitats. Kelly and Todd (1988) proposed a model for Paleoindian colonization of North America. Their predictions can be tested against the data from Charlie Lake Cave where the bison bones at the bottom of the Charlie Lake Cave deposits may well have been hunted by the first generation of people to inhabit the region in post-glacial times.

Prediction 1. Hunting should be important. Kelly and Todd argue for a primary role for hunting because it involves a set of techniques which can be transferred readily from one region to another, and because animals are available throughout the year. In more northern areas this does not apply solely to Paleoindians - hunting was of importance throughout prehistory. Bison hunting at Charlie Lake Cave is well represented, but other subsistence activities did not replace big game hunting in much of the western interior.

Prediction 2. There should be evidence for high mobility, especially in times of rapid environmental change. With only one site, this is difficult to assess. However, the lack of hearths and the discard of potentially useful artifacts (especially larger quartzite chopping tools) suggests mobility was important.

Prediction 3. Sites should be used repeatedly, but for short periods. It is difficult to assess how many times Charlie Lake Cave was visited during early post-glacial times. The excavated portion of the site which reaches the lowest components (about 12 square metres) has produced just over 100 bison bone fragments, 17 stone artifacts or cores, and less than 200 pieces of debitage (most of which derives from a couple of instances of biface resharpening). The most conservative evaluation of dates would have this deposition occurring over 500 radiocarbon years. This seems to be a series of minimal events, even if only two or three occupations occurred, and the site conforms to the prediction.

Prediction 4. Unique landscape features should be relatively unused. As Charlie Lake Cave is the only dated early post-glacial site for hundreds of kilometres in any direction, this is difficult to assess. It is clearly a unique landscape feature, but that was what attracted archaeologists there in the first place ! I disagree with this prediction. I suspect that hunter-gatherers moving into a new landscape were attracted to unique features for both practical and metaphysical reasons. Unique features help one navigate in new terrain, but are also important in the development of cognitive and spiritual maps. I have argued elsewhere that Charlie Lake Cave was 'memorialised' by its early occupants, and that deliberate burial of ravens took place there (Driver 1999).

Prediction 5. Technology should be easily transportable. As noted above, larger artifacts were discarded at the site, so transportability was probably a concern.

Prediction 6. There should be no long-term storage of food. Although the bison bones are frequently broken (presumably for marrow extraction), there is no evidence for intensive smashing of long bone epiphyses or diaphyses. Furthermore, in a relatively small assemblage of bison specimens there are a number which were not broken, including a humerus (subsequently chewed by carnivores), tibia, radius, and numerous phalanges. This contrasts with later prehistoric sites where bone smashing and boiling was common, presumably to produce fat for use in pemmican production (Brink et al. 1986; Reeves 1990). The absence of hearths also suggests minimal on-site processing. Bison bones left at the site in an unprocessed state include long bone ends which Brink (1997) ranks highly in terms of percentage of fat content. The six highest ranked long bone ends are proximal tibia, proximal humerus, proximal femur, proximal radius/ulna, distal femur and distal radius/ulna. There are no femora at the site, but a count of the minimum number of long bone ends shows that nine out of eighteen specimens are in the high-ranked bones defined above. It seems unlikely that high fat bones were being removed for processing off-site.

Discussion and conclusions

There is currently no evidence to suggest that people lived in most of what is now the western interior of Canada at any time prior to about 11,000 BP. The absence of archaeological evidence may be due to the reconfiguration of landscapes caused by the processes of deglaciation. However, palynology and vertebrate paleontology support the conclusion that this region could not support human populations until the re-establishment of vegetation (perhaps as early as 12,000 BP) and the immigration of animal populations (possibly not until 11,500 BP in the Calgary area, 11,000 BP near Edmonton, and 10,500 BP in the Peace River). The lack of late Pleistocene megafauna in post-glacial deposits suggests that most of the region was colonized by animal populations after the late Pleistocene extinctions of about 11,000 BP.

The first humans to enter the region depended on hunting, although they presumably would have gathered berries at the appropriate season. The environment they colonized was open, with areas of herbaceous brush and stands of decisuous trees. Evidence from Vermilion Lakes and Charlie Lake Cave shows that they hunted bison, with the addition of bighorn sheep and caribou closer to the mountains. By 10,000 BP much of the open landscape had been taken over by coniferous forests dominated by spruce. The only archaeological site which records this transition in any detail is Charlie Lake Cave. Vertebrates show a brief period (c. 10,500 to 10,000 BP) when animals adapted to open environments flourished, apparently at a time when aquatic habitats had yet to become sufficiently productive to attract waterfowl, beaver and muskrat. By about 10,000 BP forest species appeared, and a wide range of aquatic birds and mammals was also present.

Early human inhabitants of Charlie Lake Cave hunted bison, and may have continued to use the site for this purpose after spruce forests had begun to develop. The site was either part of a bison kill site, or very close to the kill area. Bison were butchered and bone marrow was extracted. However, not all bones were smashed to obtain marrow, and there was no further processing of bones for grease. This approach to animal exploitation, coupled with a high discard rate of formed artifacts, suggests a high mobility strategy, as predicted by Kelly and Todd (1988) for Paleoindians moving into uninhabited areas. Visits to the site ceased at about 9500 BP; the site was not re-used for another 2000 years. Later occupations were more intensive and a greater variety of activities took place (Handly 1993). It is likely that the role of the site in the settlement system changed from late-glacial to mid-Holocene times.

The early post-glacial activities at Charlie Lake Cave can be seen as a microcosm of what was probably happening over a much larger area - colonization of open, productive landscapes by high mobility hunter-gatherers; a readjustment to more forested environments with a greater diversity of fauna. However, the lack of comparative material from other early sites handicaps our understanding of the process of colonization, and the lack of early Holocene cultural components at Charlie Lake Cave means that we cannot assess the way in which human groups adapted to the arrival of coniferous forests.

References

Beaudoin, A.B., 1993. A compendium and evaluation of post-glacial pollen records in Alberta. *Canadian Journal of Archazology* 17, 92-112.

Borden, C.E., 1952. A uniform site designation scheme for Canada. *Anthropology in British Columbia* 3, 44-48.

Brink, J.W., 1997. Fat content in leg bones of *Bison bison*, and applications to archaeology. *Journal of Archaeological Science* 24, 259-274.

Brink, J.W., M. Wright, B. Dawe and D. Glaum, 1986. *Final Report of the 1984 Season at Head-Smashed-In Buffalo Jump, Alberta*. Edmonton:

Archaeological Survey of Alberta Manuscript Series 9.

Burns, J.A., 1996. Vertebrate paleontology and the alleged ice-free corridor: the meat of the matter. *Quaternary International* 32, 107-112.

Driver, J.C., 1998a. Human adaptation at the Pleistocene/Holocene boundary in western Canada, 11,000 to 9,000 BP. *Quaternary International* 49/50, 141-150.

Driver, J.C., 1998b. Late Pleistocene collared lemming (*Dicrostonyx torquatus*) from northeastern British Columbia, Canada. *Journal of Vertebrate Paleontology* 18(4), 816-818.

Driver, J.C., 1999. Raven skeletons from Paleoindian contexts, Charlie Lake Cave, British Columbia. *American Antiquity* 64(2), 289-298.

Driver, J.C., M. Handly, K.R. Fladmark, D.E.Nelson, G.M. Sullivan and R. Preston, 1996. Stratigraphy, radiocarbon dating, and culture history of Charlie Lake Cave, British Columbia. *Arctic* 49(3), 265-277.

Fedje, D.W., J.M. White, M.C. Wilson, D.E. Nelson, J.S. Vogel, and J.R. Southon, 1995. Vermilion Lakes Site : adaptations and environments in the Canadian Rockies during the latest Pleistocene and early Holocene. *American Antiquity* 60, 81-108.

Fladmark, K.R., J.C. Driver and D. Alexander, 1988. The Palaeoindian component at Charlie Lake Cave (HbRf 39), British Columbia. *American Antiquity* 53(2), 371-384.

Handly, M.J., 1993. Lithic assemblage variability at Charlie Lake Cave (HbRf-39) : a stratified rockshelter in northeastern British Columbia. Unpublished M.A. thesis, Department of Anthropology, Trent University, Peterborough, Ontario.

Kelly, R.L. and L.C. Todd, 1988. Coming into the country: early Paleoindian hunting and mobility. *American Antiquity* 53(2), 231-244.

Lichti-Federovich, S., 1970. The pollen stratigraphy of a dated section of Late Pleistocene lake sediment from central Alberta. *Canadian Journal of Earth Science* 7, 938-945.

MacDonald, G.M. , 1987. Postglacial development of the subalpine-boreal transition forest of western Canada. *Journal of Ecology* 75, 303-320.

MacDonald, G.M., R.P. Beukens and W.E.Kieser, 1991. Radiocarbon dating of limnic sediments: a comparative discussion and analysis. *Ecology* 72, 1150-1155.

MacDonald, G.M., R.P. Beukens, W.E.Kieser and D,H, Vitt, 1987. Comparative radiocarbon dating of terrestrial plant macrofossils and aquatic moss from the "ice-free corridor" of western Canada. *Geology* 15, 837-840.

MacDonald, G.M. and T.K. McLeod, 1996. The Holocene closing of the 'ice-free' corridor: a biogeographical perspective. *Quaternary International* 32, 87-95.

Mandryk, C.A.S., 1996. Late Wisconsinan deglaciation of Alberta: processes and paleogeography. *Quaternary International* 32, 79-85.

Mathewes, R.W., 1994. Evidence for Younger Dryas age cooling on the north Pacific coast of North America. *Quaternary Science Reviews* 12, 321-331.

Morlan, R., 1999. *Canadian Archaeological Radiocarbon Database*. Canadian Archaeological Association: http://www.canadianarchaeology.com.

Reeves, B.O.K., 1990. Communal bison hunters of the Northern Plains. In *Hunters of the Recent Past*, ed. L.B. Davis and B.O.K. Reeves. London: Unwin Hyman, pp.168-194.

White, J.M. and R.W. Mathewes, 1986. Postglaical vegetation and climatic change in the upper Peace River district, Alberta. *Canadian Journal of Botany* 64, 2305-2318.

Wilson, M.C., 1993. Radiocarbon dating of the ice-free corridor: problems and implications. In *The Palliser Triangle*, ed. R.W. Barendregt, M.C. Wilson and F.J. Jankunis. Lethbridge: University of Lethbridge, pp. 166-206.

Wilson, M.C., 1996. Late Quaternary vertebrates and the opening of the ice-free corridor, with special reference to the Genus *Bison. Quaternary International* 32, 97-105.

www.ingramcontent.com/pod-product-compliance
Lightning Source LLC
Chambersburg PA
CBHW061302270326
41932CB00029B/3437